Bridging the Gap to University Mathematics

Martin Gould · Edward Hurst

Bridging the Gap
to University Mathematics

 Springer

1005783751

ISBN: 978-1-84800-289-0   e-ISBN: 978-1-84800-290-6
DOI: 10.1007/978-1-84800-290-6

British Library Cataloguing in Publication Data
A catalogue record for this book is available from the British Library

Library of Congress Control Number: 2008942154

Mathematics Subject Classification (2000): 00–01, 00A05, 00A35, 00A99, 03-00, 03-01, 97-00, 97-01

Printed on acid-free paper

Springer Science+Business Media
springer.com

Preface

Mathematics has always been an exciting challenge for both of us. Even before university, we thoroughly enjoyed getting to grips with calculus and meeting stuff like vectors and mechanics for the first time. The only problem that we both found when we first started our degrees was that amongst our new peers, different people had very different mathematical backgrounds. Some had done "single maths," some "double," and some an entirely different syllabus altogether. We quickly learned that a mathematics degree requires mastery of a whole range of ideas: including those that you haven't studied before.

The aim of this book is in no way to teach you everything that you would learn in the first year of a mathematics degree. Instead, our aim was to write a book that you could read before going to university that would give you a solid foundation on which to build all of the new skills that you will acquire. That way, when you actually arrive at university you will have much more time for all of the other amazing things that being a student offers, rather than having to spend hours looking up something that you could easily learn in a few minutes from a straightforward book like this. To this end, we have included an appendix of loads of formulae and identities so that you can spend your nights partying rather than searching in the library for the integral of $\tan x$.

I suppose being young means being radical, so we've written this book backwards. Not crazy, "read in the mirror" backwards, but the chapters are set out in the reverse of what you are probably used to seeing. Each chapter is designed to be a completely stand-alone entity, and chapters always *start* with some questions. Our reason for doing this is so that if you see a chapter title about something with which you are familiar, you can dive straight into some questions then head off to the next chapter without having to read any explanation of the topic. If you really want to fly through the material, the first ten "Test Yourself" questions of each chapter are designed to cover the key points. If you can score close to full marks on these, you're doing pretty well. If what the questions are asking looks foreign to you, work through the chapter and the rest of the exercises and come back to these last. All being well, you'll be able to do them within a reasonably short time. If you have already studied a lot of "pre-university" maths, you may well have a good knowledge of quite a few of the earlier chapters' contents. If you haven't, there will obviously be more chapters where you'll be starting afresh. For those readers taking the International Baccalaureate (or any other equivalent qualification) the key to success is exactly the same – study what you need to, pass over what you don't. In any case, the latter chapters move further away from school and college mathematics and towards degree-style thinking, so as the book progresses there is certainly something for everyone.

Please don't ever be disheartened if you're finding some things difficult. We wrote this book in the hope that it could help prepare you as fully as possible for your studies, so that you can have the best time at university, both in terms of academic achievement and your student life as a whole. There may be times when you would like more practice with a new skill, and so we thoroughly recommend that in these instances you search out some more questions to do, either from the Internet or from other books.

On the whole, we hope that regardless of whether you are a student who has just decided to study maths further, or you are someone just about to set off to university, this book will serve you well as a single, cohesive guide that draws on your knowledge thus far and helps shape it so that you are ready to tackle the challenges of the mathematics ahead.

Martin Gould and Edward Hurst

Contents

Acknowledgements

This project would never have been possible without the continued support of our friends and families, to whom we wish to extend our sincerest thanks. We would especially like to thank Hannah Kimberley for her fantastic, diligent work in checking and rechecking everything we did; Professor Ian Stewart for his expert knowledge in the field; Giulian Ciccantelli and Richard Revill for their mathematical and artistic insights; Joseph Carter, Masoumeh Dashti and Hannah Mitchell for their input at the crucial collaboration phase; and our parents Tina and James Gould and Helen and Tim Hurst for putting up with a summer of us writing all the time – and eating all their food. We couldn't have done it without each and every one of you.

1
Inequalities

Test Yourself

If you think you are already comfortable with this material, try these questions first and mark them using the answers at the back of the book. If you get them all right, you're probably ready to move straight on to the next chapter. If some look tricky, study the chapter first and then come back to these when you're ready.

1. Write a list of all the whole number solutions to $5 \leq x < 8$.

2. Solve the inequality $x + 2 < 7$.

3. Solve the inequality $3x + 4 \geq 5x + 2$.

4. Solve the inequality $-3x < -12$.

5. Use a graphical method to solve $3x \geq 5 - 2x$.

6. Use a graphical method to solve $x^2 > 1$.

7. Use a graphical method to solve $x^2 - 4 < 3x$.

8. Algebraically solve the inequality $x^2 > 3x$.

9. Algebraically solve the inequality $\frac{x-4}{x+10} < 0$.

10. Algebraically solve the inequality $\frac{x+3}{x-2} < x + 6$.

E. Hurst and M. Gould, *Bridging the Gap to University Mathematics*,
DOI: 10.1007/978-1-84800-290-6_1,
© Springer-Verlag London Limited 2009

1.1 What Are Inequalities?

An inequality is just a statement that involves at least one of the signs $<$, $\leq$, $>$ or $\geq$ (i.e., "less than," "less than or equal to," "greater than," or "greater than or equal to"). In many ways we can treat them just the same as we treat equalities [i.e., any statement with the $=$ ("equals") sign], but there are some pitfalls that we need to be careful to avoid along the way, and sometimes a few extra steps are required in our calculations.

What's the Point?

Inequalities are very useful when the solution to a problem that we are considering contains lots of (perhaps infinitely many) values. Here's a really basic example:

> You have £10 in your wallet and you want to buy a CD. What price can the CD be so that you can afford it?

Clearly, the answer is anything less than or equal to £10. If we assume that the cheapest the CD could possibly be is free (so the shop aren't going to pay you to take it away from them!), then any value greater than or equal to £0.00 and less than or equal to £10.00 is OK for us. Therefore, our solution could be written as a list: £0.00, £0.01, £0.02, £0.03 and so on, up to £9.99, £10.00. The only problem with this is that writing out all these values would take the best part of a day, and the shop would be closed before you had a chance to make it to the counter with your selection. Therefore, we use an inequality to express the solution like this:

> We can buy the CD for all x satisfying: £0.00 $\leq x \leq$ £10.00.

The example of prices in a shop is a good starting point, because we can see the equivalence between the two ways of writing the answer. But what if we were thinking about some quantity where the interval (or "step") between the possible values is not so clear? With the CD, we could write out the list because we know that the price could only take values that increase by a penny at a time. How about if we were in a lab, trying to measure a temperature? There is no limit to how accurate we can go, because we could measure the temperature to 1 decimal place, to 10 decimal places, to 100 decimal places, and so on. In this instance, we could never write out all the possible values that the temperature could take – and this is where inequalities become most useful.

In the CD example, prices are what is known as *discrete*. This means they can only take certain values. In many real-life examples, such as the temperature example, the set we are working with is *continuous* – that is, it can take any value in a given range and there is no limit on the accuracy that we can measure to.

Manipulating Inequalities

Much like working with equations, the process of finding the solution set of an inequality is called "solving" it. In working with equations we know that as long as what we do to both sides of the equals sign is the same, then we will preserve the equality. The same thing is *almost* true for inequalities, but there is an important exception that we need to remember. Firstly, let's look at the things we *can* do:

- Adding (or subtracting) the same number to each side of an inequality preserves the inequality.

  For example, $x + 3 > 10$ is the same as $x > 7$; and $x - 3 > 10$ is the same as $x > 13$.

- Multiplying (or dividing) each side of an inequality by a *positive* number preserves the inequality.

  For example, $x > 6$ is the same as $2x > 12$; and $2x > 6$ is the same as $x > 3$.

- Inequalities are transitive.

  You may have not met this word before, so rather than giving a formal definition we'll just look at an "obvious" example: in an equality, if $x = 2y$ and $2y = 3z$, then $x = 3z$. The same is true if we replace equals by any of the inequality signs: if $x < 2y$ and $2y < 3z$, then $x < 3z$.

Now for the annoying thing that *doesn't* work the same as equalities:

- Multiplying (or dividing) each side of an inequality by a *negative* number <u>REVERSES</u> the inequality.

  For example, $x > -2$ is the same as $-x < 2$, and $-2x \geq 14$ is the same as $x \leq -7$.

With this last rule comes a difficulty: before we multiply (or divide) through by something, we need to be *certain* as to whether it is positive or negative. Look at this example:

  Assuming that $y \neq 0$, solve the following inequality for x: $yx < 7y$.

Gut instinct simply says $x < 7$, and as long as y can only be positive this is true. But if y is negative the solution is actually $x > 7$, because when dividing through by y we need to reverse the inequality. If you're at all unsure of how to proceed, then it's far safer to stick to addition and subtraction wherever possible when dealing with inequalities: when confronted with something like $2 > -x$, if we subtract 2 from each side we arrive at $0 > -x - 2$. Then adding x to each side gives $x > -2$: the same answer that *could* have been found by multiplying through the original inequality by -1, but without any danger of forgetting to reverse the inequality.

Now that we have these tools at our disposal, it is actually possible to solve a large number of inequalities. Here's the first set of exercises:

EXERCISES

1.1.1. Write the list of whole numbers that satisfy the inequality $7 < x \leq 10$.

1.1.2. Solve $2x > 22$.

1.1.3. Solve $x + 4 > 11$.

1.1.4. Solve $2x + 5 \leq 15$.

1.1.5. Solve $3x - 4 < x + 6$.

1.1.6. Solve $x - 7 \geq 2x + 3$.

1.1.7. Solve $3x - 5 - 4x > 7$.

1.1.8. Solve $\frac{-x}{2} + 3 \geq 3x - 4$.

1.1.9. Given that y is always positive, solve $3xy + 4y > 2y$ for x.

1.1.10. Given that y is always positive and z is always negative, solve $2xyz + yz > 3xyz$ for x.

1.2 Using Graphs

Graphical Solutions

Consider the inequality $x^2 - 4 \geq 0$. Using the basic tools that we have above, we can't do much with this. We could certainly write $x^2 \geq 4$, but where do we go from here? Before looking at the algebraic way to solve these types of inequalities, let's first look at what is actually happening by taking a graphical approach.

First of all, let's examine Figure 1.1, the graph of $y = x^2 - 4$.

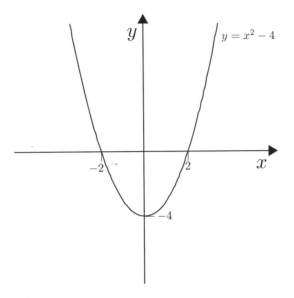

Figure 1.1

How can we use this to get our solutions? If we look at our inequality, our solutions are anything where the curve we've just drawn lies on or above the line $y = 0$ (i.e., the x-axis). So we can see that in solving this inequality, we actually have two regions where the solution is valid: $x \geq 2$ and $x \leq -2$. This might seem unusual at first, but imagine that we were solving the equality $x^2 - 4 = 0$. Here, we would have two solutions (2 and -2), so in reality what we get with the inequalities isn't that much different from what we're used to. It's worth noting, though, that if we were solving the inequality $x^2 - 4 \leq 0$, we would actually only have one *region* that served as our solution, but it would have two endpoints: this solution is just anywhere that our curve lies *on or below* $y = 0$, and so the solution is simply $-2 \leq x \leq 2$.

Of course, we can now extend this tool to solve "harder" inequalities. Consider the inequality $x^2 < x$. We now have a function in x on both sides of the inequality: but *remember what we found at the end of Section 1*: we don't know whether x is positive or negative, so we can't divide through by it! Instead, let's take the graphical approach. Draw $y = x^2$ and $y = x$ (these correspond to the two sides of the inequality), as shown in Figure 1.2.

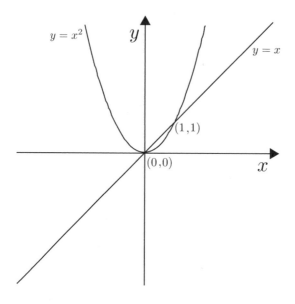

Figure 1.2

All we need to do is find the region where the x^2 curve is below the line $y = x$. We can immediately see by eye that the solution is therefore $0 < x < 1$.

Empty Solution Sets

Note that not all inequalities will have a solution. Here's an example:

$$\text{Solve } x^2 < -5.$$

Drawing $y = x^2$ and $y = -5$ very quickly reveals that the x^2 curve will never, ever lie below the line $y = -5$. In these cases, we just write "no solutions".

Now for some graph drawing of your own:

EXERCISES

1.2.1. Using a graphical method, solve $x + 2 < 7$. You learned how to solve this type of inequality in Section 1.1, so you can go back and check your answer using algebra!

1.2.2. Using a graphical method, solve $x^2 \geq 9$.

1.2.3. Using a graphical method, solve $x^2 - 100 < 0$.

1.2.4. Using a graphical method, solve $x^2 - 1 > 15$.

1.2.5. Using a graphical method, solve $x^4 < -2$.

1.2.6. Using a graphical method, solve $2x + 4 < -x$. Again, you learned how to solve this type of inequality in Section 1.1, so you can go back and check your answer using algebra!

1.2.7. Using a graphical method, solve $x^2 < 4x$.

1.2.8. Using a graphical method, solve $2x^2 \geq 18x$.

1.3 Critical Values

The final tool that we'll look at with inequalities actually requires that we treat them as equalities. This is an algebraic method for solving them, so it gives us more accurate results than drawing a graph when dealing with difficult numbers.

Finding Intersections

In drawing the graphs of the inequalities, we found that the heart of the problem was simply to find the places where the two curves or lines intersected and then to choose the regions of the graph (around these points) that satisfy the condition that we were looking for. If we can do this graphically, there is no good reason why we can't use the same idea algebraically. We call such a method "finding critical values" – that is, finding the values where the curves or lines intersect.

Every time we pass through a critical value, we know that the two curves or lines have crossed: whatever was "on top" before is now underneath, and vice versa. To clarify this, let's look again at the example $x^2 < x$. First, we solve the equation as though it were an equality, to find the critical values. We see that the critical values are 0 and 1. Now look at the region where x is less than 0. Here, the $y = x^2$ curve is higher up than the line $y = x$. Now look at any point between 0 and 1. Here, the line $y = x$ is higher than the curve $y = x^2$. Finally, look anywhere after $x = 1$, and we see that $y = x^2$ is higher than $y = x$. Every time that we pass through a critical value, we swap which is above the other. Note that the phrase "passing through" is crucial here. If we simply "touch" a critical value, it's possible that the curve might turn around and go back the way it came. For example, consider the line $y = 0$ and the curve $y = x^2$. At $x = 0$ they touch, but because one does not go *right through* the other, $y = x^2$ remains above $y = 0$. A quick and easy test to see whether lines or curves go through each other or simply touch is to substitute a value of x into each of the equations just before and just after the point that you are testing, and to look whether they have indeed "swapped around."

So all we need to do is examine any point between the critical values, and see if it satisfies our conditions. If it does, we include that region in our solution, if it doesn't, then the next region along must be in our solution instead. Here's a worked example:

$$\text{Solve } x^3 > 4x.$$

First and foremost, we have to fight our instincts to divide through by x, because we don't know whether x is positive or negative. When we've done that, we can proceed. In these calculations, it makes it easier to have a 0 on one side, so firstly let's rearrange to get $x^3 - 4x > 0$.

Now, treat it as if it's an equality, and solve it to find the critical values. The equation $x^3 - 4x = x(x + 2)(x - 2) = 0$ has three solutions, which are -2, 0 and 2. So the critical values are -2, 0 and 2.

Take the first region, which is where $x < -2$. The number -10 is good enough for us, as that's certainly less than -2. Setting x as -10, we get $-1000 + 40 = -960$. This is definitely less than 0, so this region is no good for us. Therefore, the next region along (which is $-2 < x < 0$) is OK for us so long as the curves pass through each other at the critical value. We can easily test this by substituting in a number that is present in this range: -1 seems an obvious candidate. Making the substitution yields 3 on the left-hand side, which is indeed greater than 0. By the same logic, the region after that (which is $0 < x < 2$) is definitely *not* OK for us if the curves pass through each other at the critical value. Testing the substitution $x = 1$ proves this to be true. Lastly, the final region (which is $x > 2$) must be OK for us if the curves pass through each other at the last critical value. Testing $x = 3$ shows us that this is true.

Therefore, the solution is $-2 < x < 0$ and $x > 2$. In doing just a simple equality calculation and then a quick substitution, we have found the whole solution set to the problem. This type of calculation is the key to all work with inequalities. If you didn't quite follow the logic, read over the example a couple more times.

Asymptotes

The last type of inequality that we're going to look at uses the same idea along with the idea of an asymptote. Hopefully you'll have met these before (even if you didn't know the name "asymptote"). An asymptote is just a line in a graph that a function gets closer and closer to, without actually touching. That sounds a bit wordy, but think of the graph of $y = \frac{1}{x}$. At values of x a tiny bit less than 0, the values of y are very, very negative. Suddenly, at values of x a tiny bit greater than 0, the values of y are very, very positive. This is because at $x = 0$, there is an asymptote in the graph, as shown in Figure 1.3.

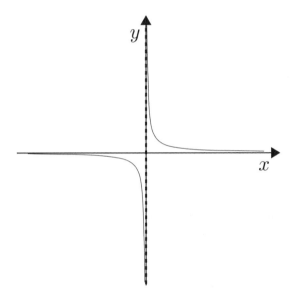

Figure 1.3

"So, how do asymptotes work with inequalities?" I hear you cry. It's reasonably simple, and here's an example:

$$\text{Solve } \frac{x+3}{x-2} < 0$$

We know from earlier examples that we need one side to be 0, but we already have this, so the first step is completed for us. On to finding the critical values: We know that if the numerator of the fraction on the left-hand side is 0, then the whole left-hand side is 0, so $x = -3$ is definitely a critical value.

Now, remember what an asymptote does: It has the potential to switch values from being positive to negative, or vice versa. That means that if a line or curve was on top before, it might be underneath after "going through" an asymptote, and vice versa. This is exactly our criteria for labelling a point as a critical value, so we know that whenever a line or curve "goes through" an asymptote, there is potentially also a critical value. Handy!

Whenever there is a 0 in the denominator we are going to have an asymptote, so all we have to do is find the values of x where the denominator is 0 and we will find the rest of the critical values. In our example, the only time that this happens is $x = 2$. Therefore, all the critical values are $x = 2$ and $x = -3$. We quickly test any point in the region $x < -3$: choosing $x = -5$ yields $\frac{2}{7}$, which is certainly greater than 0, so this region is no good for us. By making substitutions from the remaining two regions, we see that the next region is OK (i.e., $-3 < x < 2$), and the final region isn't. The solution is $-3 < x < 2$. Done!

Nonzero Woes

The only thing that could have made this problem harder was if we didn't start with a 0 on the right-hand side. Let's take a look at one final example to end the chapter:

$$\text{Solve } \frac{x + 16}{x + 4} > x - 6$$

Hopefully any temptations you have to multiply through by $x + 4$ will have long gone, so instead let's subtract $x - 6$ from both sides. This leaves us with:

$$\frac{x + 16}{x + 4} - (x - 6) > 0$$

Now, in order to do any of the tricks we've seen above, we need to combine the left-hand side to a single fraction. The common denominator of $x + 4$ looks as good as any, so let's go:

$$\frac{x + 16 - (x - 6)(x + 4)}{x + 4} > 0$$

Simplifying:

$$\frac{-x^2 + 3x + 40}{x + 4} > 0$$

Get rid of that nasty minus sign by multiplying through by -1 (remembering to reverse the inequality):

$$\frac{x^2 - 3x - 40}{x + 4} < 0$$

Now set the numerator equal to 0 and solve (we can factorise here, but sometimes we might need to make use of the quadratic formula) to get two roots of $x = -5$ and 8. Set the denominator equal to 0 and solve to get the root $x = -4$. Now examine something in the first region of $x < -5$. The value $x = -10$ is good enough for this, and it yields -5, which is indeed < 0, so this region is OK for us. The next region $(-5 < x < -4)$ is not, and the region after that one $(-4 < x < 8)$ is also OK, but the final region $(x > 8)$ isn't. Hence our solution is $x < -5$ and $-4 < x < 8$.

Here comes the final set of exercises:

EXERCISES

1.3.1. Solve $x^2 > 4$ without using a graph.

1.3.2. Solve $x^2 > 2x$ without using a graph (and place yourself under citizen's arrest if you even thought about dividing by x).

1.3.3. Solve $x^2 < 7x$ without using a graph.

1.3.4. Solve $x^2 + 3 > -4x$ without using a graph.

1.3.5. Solve $x^2 + 6x < 3$ without using a graph.

1.3.6. Solve $\frac{x+8}{x+4} > 0$ without using a graph.

1.3.7. Solve $\frac{x+6}{x-3} < 0$ without using a graph.

1.3.8. Solve $\frac{x+5}{x-2} > x + 5$ without using a graph (Division is still highly illegal: What if x were, say, -10? Who'd be laughing then?)

1.3.9. Solve $\frac{x-2}{3-x} < 3$ without using a graph.

1.3.10. Solve $\frac{2x+5}{x-3} < x + 1$ without using a graph.

Where Now?

Note that in the final section we've only dealt with strict (i.e., $<$ or $>$) inequalities. In the "critical value" method, the way we have to do things is to treat everything as though it is a strict inequality, and then at the very end manually "test" (i.e., substitute in) the endpoints of our regions to see whether they are OK or not. If they are, we use $\leq$ or $\geq$ in our solution. If not, we can only use the strict inequalities instead.

There will be plenty of time to practice manipulating inequalities at university, but it's definitely worth taking a look at some harder questions before you go. *A Concise Introduction to Pure Mathematics* (M. Liebeck, Chapman and Hall/CRC, 2000) has an excellent chapter on the topic, and some challenging exercises too.

Inequalities are used all the time in proofs. Imagine that we can prove that something is ≤ 10 and *at the same time* ≥ 10. Then the only possible value that this can take is exactly 10. This sort of logic is very helpful in analysis.

Inequalities are also very useful in disproving things. If we can show that something is both > 10 and < 10 at the same time, we know we have found an inconsistency in the original statement.

2

Trigonometry, Differentiation and Exponents

Test Yourself

If you think you are already comfortable with this material, try these questions first and mark them using the answers at the back of the book. If you get them all right, you're probably ready to move straight on to the next chapter. If some look tricky, study the chapter first and then come back to these when you're ready.

1. Use the identity $\sin^2 x + \cos^2 x = 1$ to show that $\mathrm{cosec}^2 x - \cot^2 x = 1$.

2. Find the *exact* value of $\sin\left(\frac{7\pi}{12}\right)$, leaving your answer as a surd.

3. Find the *exact* value of $\cos\left(\frac{5\pi}{6}\right)$, leaving your answer as a surd.

4. Show that $\tan(2x) = \frac{2\tan x}{\sec^2 x - 2\tan^2 x}$.

5. Use the quotient rule to evaluate $\frac{d}{dx}\left(x^2\mathrm{cosec}\ x\right)$.

6. Use the chain rule to evaluate $\frac{d}{dx}\left(\cos^2(2x)\right)$.

7. Find $\int_0^{\frac{\pi}{3}}\left(\cos^2 x - \frac{1}{2}\right)dx$.

8. Simplify $e^{7x} \cdot e^{3x}$.

9. Evaluate $\frac{d}{dx}\left(4xe^{2x}\right)$.

E. Hurst and M. Gould, *Bridging the Gap to University Mathematics*,
DOI: 10.1007/978-1-84800-290-6_2,
© Springer-Verlag London Limited 2009

10. Evaluate $\frac{d}{dx}\left(\ln(3x^2)\right)$.

2.1 Some Identities

Most of you will have met trigonometry many, many times before. This chapter is designed to be a quick whistlestop tour through lots of key ideas that are *absolutely fundamental* to mathematics: At university you won't be using these tools and concepts on their own, but rather as a small part of far wider reaching problems. As such, it's *crucial* that you're entirely comfortable with the material in this chapter. It will come up many times (without further explanation) in the course of this book!

This and That

One of the most important skills to master in trigonometry is being able to spot links between the different trigonometric functions. This lets us rewrite expressions that are difficult to work with in a "nicer," but equivalent, form. For example, you'll almost certainly know from your previous studies that $\tan x = \frac{\sin x}{\cos x}$, and often making this substitution enables us to progress through a problem. For example, if we were asked to simplify $\frac{\sin x}{\tan x}$, then by making the substitution we can immediately see that the expression cancels down to $\cos x$, eliminating both the difficulty of working with fractions and the difficulty of working with different trigonometric functions simultaneously.

Let's take a look at three more expressions which are hopefully familiar to you:

$$\mathrm{cosec}\, x = \frac{1}{\sin x}$$
$$\sec x = \frac{1}{\cos x}$$
$$\cot x = \frac{1}{\tan x} = \frac{\cos x}{\sin x}$$

If you haven't met these functions before, don't despair: there isn't really anything to get to grips with. As mathematicians, we're so familiar with the standard trigonometric functions that whenever we get confronted with a cosec, sec or cot expression, we can simply make the substitution back to sin, cos or tan if that's easier. If you're having trouble remembering which of cosec, sec and cot correspond to sin, cos and tan, the trick to remember is that the *third* letter of cosec, sec and cot will give you the *first* letter of the corresponding sin, cos or tan function.

The Big Daddy

You'll have no doubt met him before. He's the maddest, baddest identity in town:

$$\sin^2 x + \cos^2 x \equiv 1$$

Although it crops up all over the place, you might not have seen a derivation of the identity, so here's a neat one. Consider the triangle shown in Figure 2.1.

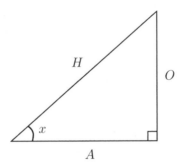

Figure 2.1

By the standard definitions of sin and cos, we arrive at $\sin x = \frac{O}{H}$ and $\cos x = \frac{A}{H}$. If we square and add these expressions, we get:

$$\sin^2 x + \cos^2 x = \frac{O^2}{H^2} + \frac{A^2}{H^2}$$
$$= \frac{O^2 + A^2}{H^2}$$

Using Pythagoras, we can see that $O^2 + A^2 = H^2$. Making this substitution into $\sin^2 x + \cos^2 x = \frac{O^2 + A^2}{H^2}$, we arrive immediately at $\sin^2 x + \cos^2 x = 1$. Sorted.

The identity $\sin^2 x + \cos^2 x = 1$ is useful on many different occasions, and from it we can even derive some other useful identities. For example, if we divide through by $\cos^2 x$, we arrive at:

$$\frac{\sin^2 x}{\cos^2 x} + \frac{\cos^2 x}{\cos^2 x} = \frac{1}{\cos^2 x}$$

$$\left(\frac{\sin x}{\cos x}\right)^2 + 1 = \left(\frac{1}{\cos x}\right)^2$$

$$\tan^2 x + 1 = \sec^2 x$$

Similarly, if we take $\sin^2 x + \cos^2 x = 1$ and instead divide through by $\sin^2 x$, we arrive at $\operatorname{cosec}^2 x = 1 + \cot^2 x$.

Angle Formulae

Here are some more very useful identities:

$$\sin(A \pm B) = \sin A \cos B \pm \sin B \cos A$$

$$\cos(A \pm B) = \cos A \cos B \mp \sin A \sin B$$

If you'd like to see a proof of one of these formulae, we provide one in "Where Now?" at the end of the chapter.

It's important to remember which way around the $\pm$ or $\mp$ signs go: in the first identity, if we're finding $\sin(A + B)$ we get $\sin A \cos B + \sin B \cos A$. In the second identity, however, the sign is *reversed:* that means $\cos(A + B) = \cos A \cos B - \sin A \sin B$. If you don't want to memorise that blindly we can actually see why it *must* be true by considering the following derivation:

- Take the equality $\cos(A + B) = \cos A \cos B - \sin A \sin B$.

- Set $B = -C$.

- Substituting this into our equality yields $\cos(A - C) = \cos A \cos(-C) - \sin A \sin(-C)$.

- Recall that sin is an *odd* function, which means that $\sin(-x) = -\sin(x)$ (if you're uncertain about this, take a look at Figure 2.2 and convince yourself!).

- Recall that cos is an *even* function, which means that $\cos(-x) = \cos(x)$ (again, see Figure 2.2 if you're not convinced).

- Using both of these facts, our identity $\cos(A - C) = \cos A \cos(-C) - \sin A \sin(-C)$ becomes $\cos(A - C) = \cos A \cos C + \sin A \sin C$.

If you're wondering what these identities are useful for, then here's an example of how they can allow us to keep our calculations *exact* (i.e., without needing to resort to a calculator and hence inevitably rounding) for as long as possible:

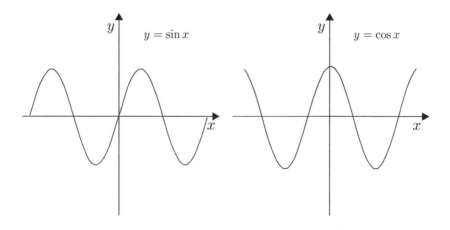

Figure 2.2

Find the *exact* value of $\cos\left(\frac{\pi}{12}\right)$.

In the trigonometric universe, it's standard to be expected to know the values of the sin, cos and tan of 0, $\frac{\pi}{6}$, $\frac{\pi}{4}$, $\frac{\pi}{3}$ and $\frac{\pi}{2}$. You *need* to know these, so get onto it now if you don't. They're in the appendix of this book if you need a reference. But how does knowing the set of angles listed above help us to find the exact value of $\cos\left(\frac{\pi}{12}\right)$? Simple – we use the identities we've just seen.

You see, $\frac{\pi}{12} = \frac{\pi}{3} - \frac{\pi}{4}$, so we can rewrite the question as: "Find the exact value of $\cos\left(\frac{\pi}{3} - \frac{\pi}{4}\right)$." From here, we're just plugging numbers into a formula:

$$\cos\left(\frac{\pi}{3} - \frac{\pi}{4}\right) = \cos\frac{\pi}{3}\cos\frac{\pi}{4} + \sin\frac{\pi}{3}\sin\frac{\pi}{4}$$

$$= \frac{1}{2}\cdot\frac{1}{\sqrt{2}} + \frac{\sqrt{3}}{2}\cdot\frac{1}{\sqrt{2}}$$

$$= \frac{1}{2\sqrt{2}} + \frac{\sqrt{3}}{2\sqrt{2}}$$

$$= \frac{1\cdot\sqrt{2} + \sqrt{3}\sqrt{2}}{2\sqrt{2}\sqrt{2}}$$

$$= \frac{\sqrt{2} + \sqrt{6}}{4}$$

And there we have it – an exact answer with no rounding.

The Double Angle Formulae

If we take the identities from the previous section and also let $A = B$, we arrive at two particularly nice identities:

$$\sin(2x) \equiv 2\sin x \cos x$$
$$\cos(2x) \equiv \cos^2 x - \sin^2 x$$

But remember "the big daddy" identity: $\sin^2 x + \cos^2 x \equiv 1$. We can rearrange this to get either of $\cos^2 x \equiv 1 - \sin^2 x$ or $\sin^2 x \equiv 1 - \cos^2 x$, and substituting these into our formula for $\cos(2x)$ yields:

- $\cos(2x) \equiv 1 - 2\sin^2 x$

- $\cos(2x) \equiv 2\cos^2 x - 1$

These little beauties come in handy when trying to integrate $\sin^2 x$ or $\cos^2 x$, because they allow us to make a clever substitution – but more about that at the end of the chapter.

Show and Tell

We'll end this section by looking at a typical question about trigonometric identities. It's fairly common to be asked to show that a given identity is true, so here's a worked example:

$$\text{Show that } \frac{\sin(2x)}{\sin^2 x} = 2\cot x$$

We're going to start by taking the left-hand side and making it into the right-hand side. Using the identity $\sin(2x) = 2\sin x \cos x$, we find that:

$$\frac{\sin(2x)}{\sin^2 x} = \frac{2\sin x \cos x}{\sin^2 x}$$
$$= 2\frac{\cos x}{\sin x}$$
$$= 2\cot x$$

Now for some exercises!

EXERCISES

2.1.1. Show that $\sec x \left(\sin^3 x + \sin x \cos^2 x \right) = \tan x$.

2.1.2. Show that $\tan x \sin x = \sec x - \cos x$.

2.1.3. Find the *exact* value of $\sin \left(\frac{\pi}{12} \right)$, leaving your answer as a surd.

2.1.4. Find the *exact* value of $\cos \left(\frac{5\pi}{12} \right)$, leaving your answer as a surd.

2.1.5. Find the *exact* value of $\sin \left(\frac{5\pi}{6} \right)$, leaving your answer as a surd.

2.1.6. Find the *exact* value of $\cos \left(\frac{7\pi}{12} \right)$, leaving your answer as a surd.

2.1.7. Express $\sin^2 x$ in terms of $\cos(2x)$.

2.1.8. Show that $\sin(2x) \tan x = 1 - \cos(2x)$.

2.1.9. Show that $\frac{\cos(2x) \sin(2x)}{2 \cos^2 x} = \sin(2x) - \tan x$.

2.1.10. Using the fact that $\tan x = \frac{\sin x}{\cos x}$, prove the following identity:

$$\tan(A \pm B) = \frac{\tan A \pm \tan B}{1 \mp \tan A \tan B}$$

2.2 Differentiating

How do we go about differentiating the basic trigonometric functions? What sort of answers should we expect? Let's start by taking a look back at the graphs of the functions $y = \sin x$ and $y = \cos x$, as shown in Figure 2.2.

We're going to take a graphical approach to this question. If we think about differentiating a function in one variable, the derivative represents the *gradient* of the function. So if we can find an explicit expression of the *gradient* of $\sin x$, we'll have found the derivative of $\sin x$. Looking carefully at the graph of $y = \sin x$, we can see that there are turning points at all of ... $\frac{-3\pi}{2}, \frac{-\pi}{2}, \frac{\pi}{2}, \frac{3\pi}{2}, \ldots$, and so all of these points have a gradient of 0. We can also examine the gradient of $y = \sin x$ at other places: if we draw the line $y = x$ on top of the curve $y = \sin x$, at $x = 0$ we see that the gradient of $y = \sin x$ at $x = 0$ is equal to 1, as shown in Figure 2.3.

Then, because we know that $\sin x$ is 2π *periodic* – that is, it repeats itself every 2π – we know that the gradient at the points ... $- 4\pi, -2\pi, 0, 2\pi, 4\pi, \ldots$ is equal to 1. We can use a similar visual argument at $x = \pi$ to see that the line there is parallel to $y = -x$, and so the gradient there is -1. We can use the periodicity again to see that the gradient at the points ... $- 3\pi, -\pi, \pi, 3\pi, \ldots$ is equal to -1.

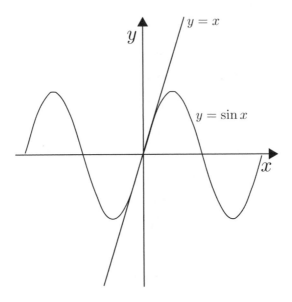

Figure 2.3

If we then find the gradient of $y = \sin x$ at various other places, we can begin to build up a picture of what the derivative of $\sin x$ will look like. Although this "visual" method is a little crude (we discuss a better method of deriving the result in "Where Now?"), it *does* provide us with the correct graphical result, as shown in Figure 2.4.

This looks strangely familiar – it's our old friend $\cos x$! We can use a similar method to derive the derivative of $\cos x$: this time, we find that the result is $-\sin x$.

Chains, Products and Quotients

Now that we have the basic results of $\frac{d}{dx}(\sin x) = \cos x$ and $\frac{d}{dx}(\cos x) = -\sin x$, we can find the derivative of a whole range of trigonometric functions. Before we do that, however, we're going to have a quick recap of the chain, product and quotient rules for differentiation. We're not going to prove them here, but there's a quick discussion of where you can find the proofs in "Where Now?" at the end of the chapter.

Let's start with the chain rule. You might not have seen it in the formal language of functions before, so here it is:

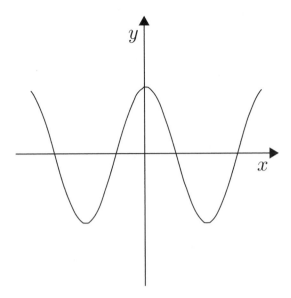

Figure 2.4

$$\frac{d}{dx}(M(N(x))) = M'(N(x)) \cdot N'(x)$$

Don't be daunted if this form of the rule isn't familiar to you. If you're used to the chain rule being written as something like $\frac{dM}{dx} = \frac{dM}{dN}\frac{dN}{dx}$, just take a moment to see that the definition above is precisely equivalent. Here's an example of the rule in action, which you should hopefully be more comfortable with. If you really *haven't* seen the chain rule before, you'll definitely need to go and look it up before proceeding.

$$\text{Evaluate } \frac{d}{dx}(\sin(2x))$$

The first thing to do is to identify how this function relates to the standard form of $M(N(x))$. We know that $M(N(x))$ is $\sin(2x)$, and so $N(x)$ must be $2x$ here. Therefore $N'(x)$, which is "the derivative of N with respect to x," is equal to 2, and $M'(N(x))$, which is "the derivative of M with respect to $N(x)$," is equal to $\cos(N(x))$. Putting everything together yields the final result $\frac{d}{dx}(\sin(2x)) = 2\cos 2x$. As we said above, if you're struggling with this you'll need to get a good explanation from somewhere, but finding one should be easy.

Next comes the product rule. This one says that, for two functions u and v in x:

$$\frac{d}{dx}(uv) = v\frac{du}{dx} + u\frac{dv}{dx}$$

We're going to jump straight in with an example:

$$\text{Evaluate } \frac{d}{dx}(\sin x \cos x)$$

We'll let $u = \sin x$ and $v = \cos x$, meaning that $\frac{du}{dx} = \cos x$ and $\frac{dv}{dx} = -\sin x$. Therefore, by the product rule, $\frac{d}{dx}(\sin x \cos x) = \cos^2 x - \sin^2 x$.

Finally, we'll look at the quotient rule. We use this to evaluate derivatives such as $\frac{d}{dx}\left(\frac{u}{v}\right)$, where u and v are functions of x. If you're happy enough using the product rule, you may be interested to note that $\frac{u}{v} = u \cdot \frac{1}{v}$. Hopefully you'll spot that this means we can actually get by *without* knowing the quotient rule specifically! However, if you're more comfortable with an explicit formula, the quotient rule is:

$$\frac{d}{dx}\left(\frac{u}{v}\right) = \frac{v\frac{du}{dx} - u\frac{dv}{dx}}{v^2}$$

As before, we won't go into a derivation, just an example:

$$\text{Evaluate } \frac{d}{dx}(\tan x)$$

Before starting to panic that we haven't worked out the derivative of $\tan x$ yet, think back to the very first thing we looked at in this chapter: the identity $\tan x = \frac{\sin x}{\cos x}$. If we make that substitution, we now know that we're actually looking to evaluate $\frac{d}{dx}\left(\frac{\sin x}{\cos x}\right)$. Call back the quotient rule, and now we're on the home straight. Letting $u = \sin x$ and $v = \cos x$, and noting that $\frac{du}{dx} = \cos x$ and $\frac{dv}{dx} = -\sin x$, we can plug everything into our quotient rule formula to arrive at:

$$\frac{d}{dx}(\tan x) = \frac{\cos^2 x + \sin^2 x}{\cos^2 x}$$
$$= \frac{1}{\cos^2 x}$$
$$= \sec^2 x$$

There's been lots of material in this section, and it's all be covered very quickly – make sure to have a thorough bash at the exercises.

EXERCISES

2.2.1. Use the product rule to evaluate $\frac{d}{dx}\left(x^2 \cos x\right)$.

2.2.2. Use the product rule to evaluate $\frac{d}{dx}\left(\frac{1}{2}x^6 \tan x\right)$.

2.2.3. Use the quotient rule to evaluate $\frac{d}{dx}\left(\mathrm{cosec}x\right)$.

2.2.4. Use the quotient rule to evaluate $\frac{d}{dx}\left(\cot x\right)$.

2.2.5. Evaluate $\frac{d}{dx}\left(x \sec x\right)$.

2.2.6. Use the chain rule to evaluate $\frac{d}{dx}\left(\frac{1}{2}\sin(4x)\right)$.

2.2.7. Use the chain rule to evaluate $\frac{d}{dx}\left(\cos^6 x\right)$.

2.2.8. Use the chain rule (twice!) to evaluate $\frac{d}{dx}\left(\sin^2\left(\frac{x}{2}\right)\right)$.

2.2.9. Evaluate $\int_0^{\frac{\pi}{4}} \cos(3x)dx$.

2.2.10. Evaluate $\int_0^{\frac{\pi}{3}} \sin^2 x \, dx$.

2.3 Exponents and Logarithms

Consider the line $y = x$. At any point on the line, the gradient is equal to 1. Nothing too special there. Let's turn the heat up a little, and consider the curve $y = x^2$. Now, the gradient at any point is equal to twice the x coordinate: what we see is a functional relationship between the x coordinate and gradient of the line.

That's all well and good, but why is it interesting to us? In both of those cases, we might consider the gradient to be a by-product of the way that we define the line or curve: Simply differentiating the equation of the line or curve gives us the result. But how about turning things on their head? How about considering something whose *derivative* was the most interesting thing about it?

Imagine a curve whose gradient was always equal to the equation of the curve itself. That would mean that for any value of x, the gradient would be equal to the value the function itself takes. Sound a little crazy? Well the exponential function e^x does exactly that. At $x = 0$, the gradient of the function is equal to $e^0 = 1$. At $x = 1$, the gradient of the function is equal to e^1, at $x = 2$, the gradient of the function is equal to e^2, and so on. But what is this elusive e?

Back to Reality

e is like π. It's a real number that has an amazing property: Whereas π helps us describe the relationship between the radius and the circumference of a circle, e helps us describe the function whose gradient is always equal to the value of the function itself. A nifty property, I'm sure you'll agree, but e is a real number, *and don't you forget it!* You can find an approximation to the value of e using a calculator, but to five decimal places $e \approx 2.71828$.

Figure 2.5 shows the graph of the curve $y = e^x$.

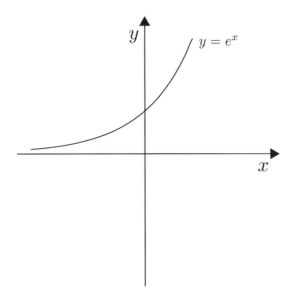

Figure 2.5

From the graph, we can note some interesting properties of the the exponential function, $y = e^x$:

- $e^0 = 1$.

- As x gets very negative, the value of e^x approaches 0, but *from above* (so e^x *never* takes a value less than 0).

- As x gets larger, e^x gets larger *incredibly quickly*.

One crucial thing to remember about e^x is the very property that defines it: The gradient is equal to the value the function itself takes. Stated formally, we write $\frac{d}{dx}\left(e^x\right) = e^x$. Not too hard to remember, is it? *The derivative of e^x is e^x.*

Remember that the standard rules for working with powers apply to e^x too, so we have:

- $e^a \cdot e^b = e^{a+b}$

- $\frac{e^a}{e^b} = e^{a-b}$

- $\left(e^a\right)^b = e^{ab}$

One last tool that often comes in handy when working with e^x is the chain rule. If we're trying to evaluate $\frac{d}{dx}\left(e^{f(x)}\right)$, the chain rule tells us that the answer we'll get is $f'(x) \cdot e^{f(x)}$. Make sure you're happy with where this comes from. Remember that the derivative of $e^{f(x)}$ *with respect to* $f(x)$ is simply $e^{f(x)}$. Here's a concrete example:

$$\text{Find } \frac{d}{dx}\left(e^{x^2}\right)$$

Using what we have above, we see that the answer is $2x \cdot e^{x^2}$. That's it!

Turn It, Leave It, Stop – Invert It

As a function, just like many of the "old favourites," e^x has an inverse. Although it has many interesting properties in its own right, we're going to think of the *natural logarithm* of x, written $\ln(x)$, as simply being the inverse of e^x. As such, we can immediately state that:

- $\ln(e^x) = x$

- $e^{\ln(x)} = x$

As with all invertible functions, we can find the graph of $\ln(x)$ by reflecting its inverse, e^x, in the line $y = x$, as shown in Figure 2.6.

Hopefully you'll be very familiar with how to manipulate logarithms in general (if you aren't, you'll definitely benefit from taking some time to research this before university), and the natural logarithm is no exception to the rules that you know and love. In fact, the natural logarithm is so closely related to e because it is the logarithm with *base* e; it's that simple! There's a list of some of the "tricks" that we can employ when working with logarithms in the appendix of this book, but one final thing that we're going to examine a little more closely is the derivative of $\ln(x)$.

We're going to attempt to discover the derivative of $\ln(x)$ in the same way that we did for $\sin x$, that is, take a look at the graph of the function and see how the gradient would look if we plotted it.

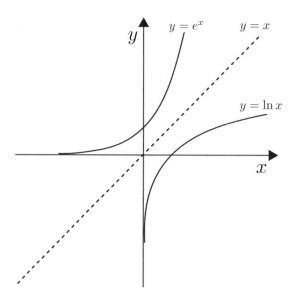

Figure 2.6

At values of x that are only slightly greater than 0, the gradient of $\ln(x)$ is very, very large. As we increase the value of x we find that the gradient of $\ln(x)$ falls rapidly, and then for very large values of x the gradient of $\ln(x)$ approaches 0 (but from above, so its gradient is never less than 0). If we carefully draw a graph of these results, we would find that it looks like Figure 2.7.

Again, hopefully you'll recognise this old chestnut: Figure 2.7 is the graph of $y = \frac{1}{x}$. Admittedly, it's restricted to the positive values of x, but then again $\ln(x)$ is *only defined* for these values of x, so we're OK! So, stating it formally: $\frac{d}{dx}(\ln x) = \frac{1}{x}$.

We'll conclude the chapter by again calling on the chain rule. You're probably sick of reading it by now, so we'll dive straight into an example and then head off to the exercises:

$$\text{Evaluate } \frac{d}{dx}(\ln(2x))$$

The answer is fairly straightforward by the chain rule, but it might not be what you expect: The derivative of $2x$ is 2, and so $\frac{d}{dx}(\ln(2x)) = 2 \cdot \frac{1}{2x} = \frac{1}{x}$. If you thought that the answer was going to be $\frac{2}{x}$ or $\frac{1}{2x}$, you just need to be very careful when using the chain rule because we get a cancellation that you might not predict.

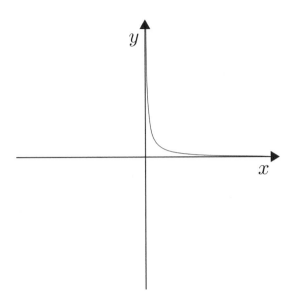

Figure 2.7

EXERCISES

2.3.1. Simplify $e^{4x} \cdot e^{3x}$.

2.3.2. Simplify $\frac{e^{6y}}{e^{3x}}$.

2.3.3. Simplify $\ln(e^{4x})$.

2.3.4. Simplify $e^{\ln(x^3)}$.

2.3.5. Evaluate $\frac{d}{dx}\left(e^{2x}\right)$.

2.3.6. Evaluate $\frac{d}{dx}\left(\frac{1}{3}e^{x^3}\right)$.

2.3.7. Use the product rule to find $\frac{d}{dx}\left(e^x \sin x\right)$.

2.3.8. Evaluate $\frac{d}{dx}\left(\ln(x^3)\right)$.

2.3.9. Evaluate $\frac{d}{dx}\left(e^{x^2}\ln(4x)\right)$.

2.3.10. Evaluate $\frac{d}{dx}\left(4x^3\ln(x^2)\right)$.

Where Now?

In this chapter we've looked at a large assortment of different tools that come in handy in all sorts of mathematical problems. To state precisely where these tools lead would require a whole book in itself, and even then would only cover the tip of the iceberg. Throughout the rest of this book you'll be using these techniques, and so you'll be able to gain some insight into where you might employ specific mathematical techniques.

In this chapter we've touched on a couple of ideas that may be aided by a lengthier explanation than was already given. If you find these sorts of proofs helpful, then we recommend you take a careful look at what follows. If you find it a bit confusing, don't be afraid to head straight to the next chapter.

We'll start with a proof of the addition formula $\sin(\alpha + \beta) = \sin\alpha\cos\beta + \sin\beta\cos\alpha$. If you enjoy this, proofs of other addition and subtraction formulae are readily available in *Analysis by its History* (E. Hairer and G. Wanner, Springer-Verlag, New York, 1996). The proofs there don't take a geometric route like we have, but they're enjoyable all the same!

We'll be using Figure 2.8 throughout the proof.

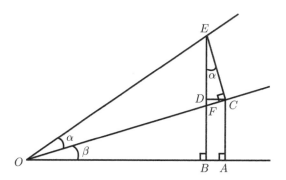

Figure 2.8

Angles $E\hat{D}C$ and $E\hat{C}O$ are right angles, as are all four angles in the rectangle $ABCD$. Here's the proof:

1. Scale the whole diagram appropriately so that OE is of length 1.

2. Using $\sin\beta = \frac{\text{opposite}}{\text{hypotenuse}}$, we get $\sin\beta = \frac{EC}{1} = EC$.

3. Using $\cos\beta = \frac{\text{adjacent}}{\text{hypotenuse}}$, we get $\cos\beta = \frac{OC}{1} = OC$.

4. Using $\sin\alpha = \frac{\text{opposite}}{\text{hypotenuse}}$, we get $\sin\alpha = \frac{AC}{OC}$.

5. Rearranging this gives $AC = OC \sin \alpha = \cos \beta \sin \alpha$, by step 3.

6. Figure 2.9 is a "zoomed in" picture of $CFDE$, and here's an explanation of why the angles are what they are:

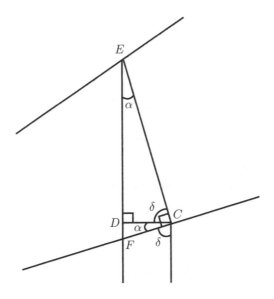

Figure 2.9

- Let $\delta = \pi - \alpha$.

- Because the angles in a triangle sum to π, we know that $O\hat{C}A = \delta$.

- Because $A\hat{C}D$ is a right angle, we know that $F\hat{C}D = \alpha$.

- Because $F\hat{C}E$ is a right angle, we know that $D\hat{C}E = \delta$.

- Because the angles in a triangle sum to π, we know that $D\hat{E}C = \alpha$.

7. Using $\cos \alpha = \frac{\text{adjacent}}{\text{hypotenuse}}$, we get $\cos \alpha = \frac{ED}{EC}$.

8. Rearranging this gives $ED = EC \cos \alpha = \sin \beta \cos \alpha$, by step 2.

9. Using $\sin(\alpha + \beta) = \frac{\text{opposite}}{\text{hypotenuse}}$, we get $\sin(\alpha + \beta) = \frac{EB}{1} = EB = ED + DB$.

10. $DB = AC$ because $ABCD$ is a rectangle.

11. Combining steps 9 and 10 yields $\sin(\alpha + \beta) = ED + AC$.

12. Combining steps 5 and 8 with step 11 yields $\sin(\alpha + \beta) = \sin \alpha \cos \beta + \sin \beta \cos \alpha$. There we have it!

For those of you with a keen eye for detail, we expect that our "visual" method of finding derivatives was a little unsatisfactory. Please don't despair: the *results* that we gave are indeed true, but there are much more rigorous ways to achieve them than we presented. One of the more accessible comes from power series, and this is discussed fully in *Real Analysis* (J. M. Howie, Springer-Verlag, New York, 2001).

In the chapter we also mentioned the fact that we can use the formulae for $\cos(2x)$ to help us integrate functions involving $\sin^2 x$ and $\cos^2 x$. Here's how:

$$\int \cos^2 x \, dx = \frac{1}{2} \int \left(\cos(2x) + 1 \right) dx$$
$$= \frac{1}{2} \left(\frac{\sin(2x)}{2} + x \right) + c.$$

By making the clever substitution, we've avoided all of the messy side calculations that are required for an integration by parts – a real time saver, we're sure you'll agree.

Finally, for those of you keen to see a derivation of the chain, product and quotient rules, we'd recommend that you dive into *Calculus I* (J. Marsden and A. Weinstein, Springer-Verlag, New York, 1999). The derivations rely on the idea of taking a limit, but if you're interested in taking a look, then the text explains everything clearly.

3
Polar Coordinates

Test Yourself

If you think you are already comfortable with this material, try these questions first and mark them using the answers at the back of the book. If you get them all right, you're probably ready to move straight on to the next chapter. If some look tricky, study the chapter first and then come back to these when you're ready.

1. Express in polar coordinates the point that is three units away from the origin and lies on the negative x-axis.

2. Express in polar coordinates the point that is 1 unit away from the origin (along the shortest path), lies on the Cartesian line $y = x$ and is in the first quadrant.

3. Consider the point $(-2, 2)$ in Cartesian coordinates. What is this point in polar coordinates?

4. Consider the point $(-1, -\sqrt{3})$ in Cartesian coordinates. What is this point in polar coordinates?

5. What is the polar form of the equation of the half-line that lies on the Cartesian line $y = x$ and is in the third quadrant?

6. What is the polar form of the equation of the circle that passes through the Cartesian points $(1, 0)$, $(3, 0)$ and $(2, 1)$?

E. Hurst and M. Gould, *Bridging the Gap to University Mathematics*,
DOI: 10.1007/978-1-84800-290-6_3,
© Springer-Verlag London Limited 2009

7. What is the polar form of the equation of the straight line passing through the Cartesian points $(-2, 0)$ and $(-2, -1)$?

8. What is the general polar equation of an Archimedean spiral?

9. Consider the Cartesian point $(2, -2, 2)$ in 3-dimensional space. What is this point in spherical polar coordinates?

10. A murder has taken place at the origin, and a police helicopter is encircling the crime scene. The helicopter maintains a constant height of 50 m above the ground, and the radius of the helicopter's circular path is 20 m. State the equation of the helicopter's path in cylindrical polar coordinates.

3.1 A Different Slant

Usually when working with curves, we use the *Cartesian coordinate* system: that is, we define a curve by the perpendicular distance from the x-axis and the perpendicular distance from the y-axis at any given point. You may also have met *parametric* curves: these are curves where the equation that defines the line is given in terms of x, y and some other parameter that varies between two specified limits. There is, however, another way.

Polar coordinates use the idea that every point in a 2-dimensional plane can be uniquely determined by just two pieces of information:

- The distance from the origin.

- The angle that the straight line joining the point to the origin makes with the positive x-axis.

Therefore, in polar coordinates we use the variables r and θ, where r is the distance that a given point is from the origin and θ is the angle that the line joining the point to the origin makes with the positive x-axis. We write polar coordinates in the form (r, θ), where θ is measured in radians unless otherwise specified. Look at Figure 3.1, which illustrates the concept, and have a think to make sure that you're totally happy with the idea – including the fact that by specifying r and θ, we *uniquely* determine a point on the plane.

Convention, Convention, Convention

When working with standard Cartesian coordinates in two dimensions, everyone in the world abides by the convention that the horizontal axis is the x-axis,

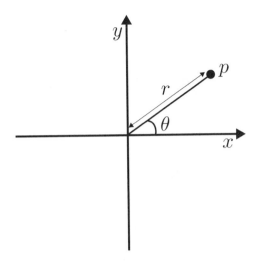

Figure 3.1

and the vertical axis is the y-axis. If we didn't universally agree on this, and some people put their axes a different way around, then different people would describe points as having different coordinates! Obviously this is no help to anyone, and so the convention has been set – and therefore must be adhered to.

Polar coordinates have their own set of conventions too. We've already met one of them – the fact that we're always going to take the angular measurement from the positive x-axis. We could, of course, choose any line as our initial point of reference, but having everyone agree that the positive x-axis is the line to be used ensures that people can agree when describing a point in polar coordinates.

Another thing to note is that angles are always measured *anticlockwise*. You can see this in action in the example above, and this convention is applicable right across mathematics. By measuring angles in this way, there is a logical method to numbering the four areas of the plane divided up by the axes. We call these areas *quadrants*, and Figure 3.2 shows how they are numbered.

This labelling is very helpful to us with polar coordinates. If we were to draw a shape and wanted to discuss only the upper left-hand region of the shape, we could easily refer to this as "the set of points in the second quadrant."

Back and Forth

When we're working with polar coordinates, we can do many of the things that we would normally do with Cartesian coordinates without needing to change

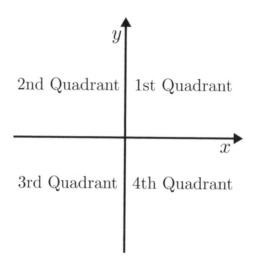

Figure 3.2

back and forth between the two different systems. There are, however, times
when having information about a point or a curve in the Cartesian form is
preferable to having it expressed in polar coordinates (or vice versa) and so we
need a way to convert between the two different systems. Such a conversion
requires expressing x and y in terms of r and θ. Thankfully, with a small bit of
trigonometry, this can be easily achieved (with a small bit of trigonometry, is
there anything that *can't* be?). Figure 3.3 shows a point, P, expressed in the
two different systems.

By drawing in those dotted lines, it makes it clear that $\cos\theta = \frac{x}{r}$, and so
$x = r\cos\theta$. Similarly, $\sin\theta = \frac{y}{r}$, and so $y = r\sin\theta$. Therefore, given r and θ,
it's very easy to calculate x and y. These conversions are *absolutely crucial*, so
commit them to memory now!

We can, of course, also convert from Cartesian coordinates to polar coordinates. From the x and y that we are given, we need to somehow determine the
distance of the point from the origin, and also the angle that the line that con-
nects it to the origin makes with the positive x-axis. Finding the distance is a
job for Pythagoras and the angle is easy when we call on tan, so this conversion
isn't difficult either. Here's a couple of worked examples to show conversion in
action, and then we're ready to tackle some exercises.

- Convert $(2\sqrt{3}, \frac{\pi}{3})$ from polar to Cartesian form.

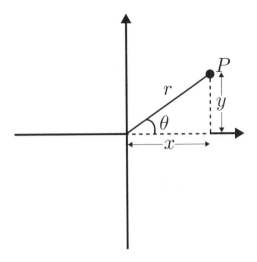

Figure 3.3

$$x = r\cos\theta$$
$$= 2\sqrt{3}\cos\frac{\pi}{3}$$
$$= \sqrt{3}$$
$$y = r\sin\theta$$
$$= 2\sqrt{3}\sin\frac{\pi}{3}$$
$$= 3$$

So the coordinates in Cartesian form are $(\sqrt{3}, 3)$.

- Convert $(-3, 3)$ from Cartesian to polar form.

  · Use Pythagoras to find the distance: $r = \sqrt{(-3)^2 + 3^2} = 3\sqrt{2}$.

  · Use tan to find the angle: this time we're working in the second quadrant, so things aren't so straightforward. If we find the angle α that the line joining the point to the origin makes with the *negative* x-axis, we can then find θ by performing $\theta = \pi - \alpha$: if you're not sure why this is the case, try drawing a diagram. So $\tan\alpha = \frac{3}{3} = 1$, so $\alpha = \frac{\pi}{4}$ and hence $\theta = \frac{3\pi}{4}$.

So the coordinates in polar form are $(3\sqrt{2}, \frac{3\pi}{4})$.

EXERCISES

3.1.1. Express the point that is two units away for the origin and lies on the positive x-axis in polar coordinates.

3.1.2. Express the point that is four units away for the origin and lies on the negative y-axis in polar coordinates.

3.1.3. Convert the point $(2, \pi)$ from polar to Cartesian coordinates.

3.1.4. Convert the point $(10, \frac{\pi}{4})$ from polar to Cartesian coordinates.

3.1.5. Convert the point $(8, \frac{\pi}{6})$ from polar to Cartesian coordinates.

3.1.6. Convert the point $(2, 2)$ from Cartesian to polar coordinates.

3.1.7. Convert the point $(0, 3)$ from Cartesian to polar coordinates.

3.1.8. Convert the point $(-\sqrt{3}, 1)$ from Cartesian to polar coordinates.

3.2 Lines and Circles

In Cartesian coordinates, describing lines and curves is probably second nature to you now, having had it drilled in from a young age. In polar coordinates, however, things are a little bit different. Without using trigonometry, we can get into some trouble when we need to describe certain lines. There are, however, some "easier" lines to look at, and so we'll befriend those first.

Straight Lines Through the Origin

When most people first meet polar coordinates, their gut instinct is to convert back to Cartesian form before attempting anything tricky. In order to fight this temptation from the outset, we're going to look at some lines that polar coordinates tackle *a darn sight better* than the Cartesian system does. These are straight lines that lie radially from the origin – often called "half lines."

In the previous section, we already got to grips with the idea that any *point* in the space can be uniquely expressed by stating its distance from the origin, and the angle that the line that joins it to the origin makes with the positive x-axis. But how about if we take away the constraint of stating the distance from the origin? Then we're just left with an angle from the positive x-axis – and this describes a line!

Let's take a look at an example. If we are given the information $\theta = \frac{3\pi}{4}$, but no value of r, then we have an infinite number of points that satisfy the condition. Drawing all of these "points," we arrive at Figure 3.4.

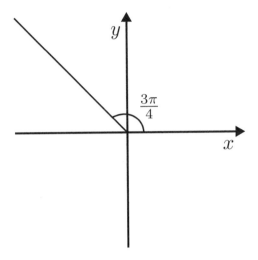

Figure 3.4

Quite impressive, I'm sure you'll agree. In Cartesian coordinates, we would have had to have described the line $y = -x$, and also the restriction that we only want the part of the line that is in the second quadrant. In polar coordinates, we can describe the same line by the simple statement $\theta = \frac{3\pi}{4}$.

More Lines

There are, of course, other straight lines in the plane. I'm sure you're dying to scream "$y = mx + c$" right now, and so let's take a look at *those* kinds of lines – lines that have some given gradient and some intercept on the y-axis, other than the origin. Well, once again thanks to some clever trigonometry, we can actually express these lines quite easily in polar coordinates too. Because the concept of "gradient" isn't really relevant in polar coordinates, where we prefer to think of things in terms of angles, the first job is to find some alternative description of our line *without* using the idea of "how much it goes up for every unit that it goes along." Thankfully, all we need to do is describe another line that our line is *perpendicular* to, and then we know what "steepness" we're going to be dealing with. But why is describing this new line, which our line is perpendicular to, any better than describing our line in the first place? Easy:

We can choose our perpendicular line to be a line that goes through the origin, which we already know how to completely describe!

If we were trying to convert the line $y = x + 6$ from Cartesian to polar coordinates, we could see that this line is definitely perpendicular to the line $y = -x$, which can be described as $\theta = \frac{3\pi}{4}$ for the second quadrant. Therefore, we know that the line we're dealing with is perpendicular to the line $\theta = \frac{3\pi}{4}$. Figure 3.5 is a diagram of the two lines, just to clarify exactly what's going on.

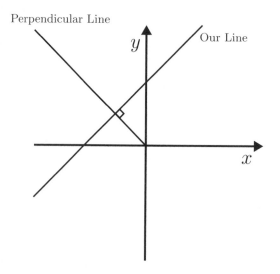

Figure 3.5

Once we have a line that our line is perpendicular to, if we also state the point at which the two lines intersect we will have uniquely determined our new line. We know how to do this already: It's simply a case of describing a single point in the plane, using polar coordinates.

Now that we have all of this information, we just need to put it together. Without worrying too much about where the equation comes from (there's an explanation in the "Where Now?" section at the end of this chapter, so when you're comfortable with the rest of the material you may want to take a look at that too!) we can state the formula:

$$r = \frac{r_0}{\cos(\theta - \theta_0)}$$

where our line is perpendicular to the line $\theta = \theta_0$, and the lines intersect at the point (r_0, θ_0).

Let's return to the example:

Find the polar equation of the line given in Cartesian form as $y = x + 6$.

Firstly, we can see that this line is perpendicular to the line $\theta = \frac{3\pi}{4}$, or $y = -x$ in Cartesian form. Next, we need to find the point where these lines intersect: it's best to proceed in Cartesian form here. The point where the lines intersect is precisely the point where the y values agree, and so finding this y is just a matter of solving $x + 6 = -x$, and so $x = -3$. Then, using $y = -x$, we see that $y = 3$, and so the point of intersection is $(-3, 3)$. In polar coordinates, this is $(3\sqrt{2}, \frac{3\pi}{4})$, using the techniques from the previous section. And there we have it: We have our (r_0, θ_0); it is the point $(3\sqrt{2}, \frac{3\pi}{4})$. Now all that's left to do is substitute this into the formula, giving $r = \frac{3\sqrt{2}}{\cos(\theta - \frac{3\pi}{4})} = 3\sqrt{2}\sec(\theta - \frac{3\pi}{4})$.

Circular Logic

In addition to straight lines, polar coordinates are excellent at describing circles too. Again, we're going to start out by looking at the "easy" case, and then move on to general circles.

Let's begin by considering a circle that is centred at the origin. When we were describing straight lines and dealing with the origin, we found that we only needed to specify a value for θ, because we could let r vary. Hopefully, right now you'll be having the dawning of realisation that if we do things the other way around (i.e., specify r and let θ vary) we get a circle of radius r, centred at the origin! It really is that easy: We simply state the radius of the circle that we require, *and we're done!* None of this grappling with x^2 and y^2 that we get with Cartesian coordinates. Just specify the radius and run. If we need a circle of radius 6 that is centred at the origin, we write $r = 6$ and then go to the bar. It's over before it even begins.

The only sad part of this story is that circles that are not centred at the origin aren't quite so quick to finish off. Not far off, but not quite. Recall that the formula in Cartesian coordinates for a circled centred at (x_0, y_0) of radius a is given by:

$$(x - x_0)^2 + (y - y_0)^2 = a^2$$

Let's try setting some polar substitutions into this formula:

- First, substitute $x = r\cos\theta$, $x_0 = r_0\cos\theta_0$, $y = r\sin\theta$ and $y_0 = r_0\sin\theta_0$:

$$(r\cos\theta - r_0\cos\theta_0)^2 + (r\sin\theta - r_0\sin\theta_0)^2 = a^2$$

- Expand the brackets on the left-hand side:

$$r^2\cos^2\theta - 2r\cdot r_0\cos\theta\cos\theta_0 + r_0{}^2\cos^2\theta_0 + r^2\sin^2\theta - 2r\cdot r_0\sin\theta\sin\theta_0 + r_0{}^2\sin^2\theta_0$$

- Factorise:

$$r^2(\sin^2\theta+\cos^2\theta)-2r\cdot r_0(\cos\theta\cos\theta_0+\sin\theta\sin\theta_0)+{r_0}^2(\sin^2\theta_0+\cos^2\theta_0)=a^2$$

- Use the identity $\sin^2\theta+\cos^2\theta=1$:

$$r^2-2r\cdot r_0(\cos\theta\cos\theta_0+\sin\theta\sin\theta_0)+{r_0}^2=a^2$$

- Use the identity $\cos(A-B)=\cos A\cos B+\sin A\sin B$:

$$r^2-2r\cdot r_0\cos(\theta-\theta_0)+{r_0}^2=a^2$$

This is the formula that we're going to be using, so commit it to memory now! If you don't let it frighten you and just set about remembering it, you'll be done within a minute. The radius of the circle is a, and its centre is the point (r_0,θ_0). Here is a worked example, before some exercises for you to show off your new skills with.

Find the equation of the circle with a radius of 2 and centre with polar coordinates $(4,\frac{\pi}{2})$.

$$r^2-2r\cdot r_0\cos(\theta-\theta_0)+r_0^2=a^2$$
$$r^2-2r\cdot4\cos\left(\theta-\frac{\pi}{2}\right)+16=4$$
$$r^2-8r\cos\left(\theta-\frac{\pi}{2}\right)+12=0$$

We can't simplify this any further. Don't fret about it – it's perfectly correct!

EXERCISES

3.2.1. What is the equation, in polar coordinates, of the half-line that starts at the origin and runs along the positive y-axis?

3.2.2. What is the equation, in polar coordinates, of the line that lies on the Cartesian line $y=-x$ and lies only in the second quadrant?

3.2.3. What is the equation, in polar coordinates, of the line that lies on the Cartesian line $y=x$ and lies only in the third quadrant?

3.2.4. What is the equation, in polar coordinates, of the circle that is centred at the origin and has a radius of 4?

3.2.5. Convert the line $y=x+1$ from Cartesian to polar form.

3.2.6. Convert the line $y=4-x$ from Cartesian to polar form.

3.2.7. What is the polar equation of the circle of radius 1 whose centre lies at the Cartesian point $(1, 0)$?

3.2.8. What is the polar equation of the circle of radius 7 whose centre lies at the Cartesian point $(1, 1)$?

3.3 Moving on Up

We've now firmly established that polar coordinates are capable of dealing with a whole range of tricks in two dimensions. But what about three dimensions? Can we use what we've worked through so far, and extend it so that we can describe 3-dimensional shapes and surfaces? You bet! In fact, polar coordinates provide us with more than one way of doing so! Let's take things up a dimension.

Cylindrical Polar Coordinates

Cylindrical polar coordinates are, unsurprisingly, based on the idea of extending the "circular" approach in two dimensions up to a "cylindrical" one in three. You see, if we wanted to describe any point in a 3-dimensional space, we could do this by imagining that we were looking at it from above, from within a cylinder. Sound weird enough for you?

The concept is actually not too hard to grasp. All that we do is describe what we see in two dimensions through the cylinder, and then specify the height of the point above the plane. If that sounds a little wordy, think about it like this. Imagine that there is a light shining directly down on an opaque point somewhere in a 3-dimensional space. Without looking at the point itself, look at the plane that the origin is in, and use standard polar coordinates to describe the location of the shadow that you see there. Because the light is shining straight down, this point is the *projection* of our point onto the plane containing the origin.

When we have this piece of information, simply describe how high up from this plane the point really is. If you've followed the logic correctly, you'll recognise that we have *uniquely* determined a point in a 3-dimensional space. We used standard polar coordinates to describe where our point lies in a circle, and then how high up this circle needs to be lifted so that the point can be found. In essence, we have visualised everything in terms of a cylinder. Cylindrical polar coordinates are stated in the form (r, θ, t), where r and θ are the standard polar parameters, and t is the height of the point above the horizontal plane containing the origin.

Figure 3.6 is a diagram to illustrate the idea. The "cylinder" is only drawn in for reference, and the thick black line on the x, y-plane is the line joining the origin to the *projection* of the point. The vertical dotted line then shows t, the height of the point itself above the plane.

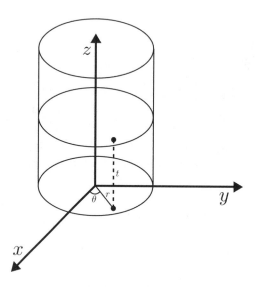

Figure 3.6

Spherical Polar Coordinates

In Cartesian coordinates, we define a point in 3-dimensional space by stating three distances (the x, y and z coordinates). In cylindrical polar coordinates, we state two distances and one angle. The other approach – that of *spherical* polar coordinates – is to give two angles and one distance.

Firstly, let's deal with the distance. In spherical polar coordinates, r is the distance from the origin to the point in the 3-dimensional space. In other words, it is the radius of a sphere, on which our point lies, that is centred at the origin. Note that this is slightly different from the r that we worked with in cylindrical polar coordinates: there, r represents the radius of the cylinder on which our point lies, and so r was found using a 2-dimensional calculation (we achieved this by taking a projection).

We find our next parameter by looking at the projection of our point onto the plane containing the origin, just like we did with cylindrical polar coordinates. We find φ: the angle that the straight line joining the projection of the point to the origin makes with the positive x-axis.

When we have these two pieces of information, however, we *don't* then state the height of the point above this plane; instead, we state the *angle* that the line joining the point to the origin makes with the *vertical* plane. See what's happening? When we're looking at the projection in the second step, we're looking at the angle made with the horizontal plane. Then, when we're looking at the actual point in the third step, we go to looking at the angle made with the *vertical* plane. We measure this parameter *downwards* from the Cartesian z-axis, and so it can only ever take a value between 0 and π. This is our third parameter, θ. It might seem a little tricky to get your head around, but after a few reads through enlightenment should dawn. Figure 3.7 should help you to visualise what's going on.

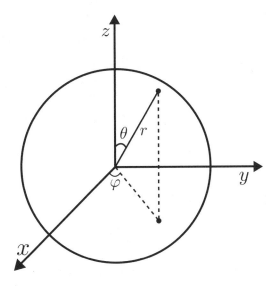

Figure 3.7

Spherical polar coordinates are stated in the form (r, θ, φ): r is the total distance from the origin to the point, θ is the angle that the line between the point and the origin makes with the *vertical* plane and φ is the angle between the line joining the projection of the point and the positive x-axis. Make sure you're careful with the *order* that r, θ and φ are stated in. It's doesn't necessarily relate to the logical order in which you work them out!

The Eye of the Beholder

So far, this chapter has been all about expressing familiar shapes in a new way. From now on, we're going to look at some *really cool stuff:* shapes that you won't have met in Cartesian coordinates. In terms of "real" mathematics, the uses of this kind of knowledge will be rather limited, but it's still definitely worth a look. Buckle up: We're going to end the chapter by taking a whistlestop tour through some of the awesome 2-dimensional shapes that can be constructed with polar coordinates. Don't worry too much about remembering all of this stuff, it's partially included for fun, because the shapes are so impressive!

- The Limaçon

  The limaçon is a curve with the polar equation $r = a + b\cos\theta$ if its line of symmetry lies along the x-axis, or $r = a + b\sin\theta$ if its line of symmetry lies along the y-axis. Figure 3.8 shows a few examples of the curve with its symmetry along the x-axis, in order to highlight how varying a varies the curves. Varying b just varies the size.

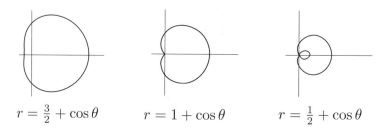

$$r = \tfrac{3}{2} + \cos\theta \qquad\qquad r = 1 + \cos\theta \qquad\qquad r = \tfrac{1}{2} + \cos\theta$$

Figure 3.8

- The Cardioid

  There is a special kind of limaçon, called the *cardioid*, which occurs when $a = b$. You'll notice that the middle of the three diagrams above is actually a cardioid, and perhaps this will give you a good idea as to why it is called what it is: It looks like a heart!

- The Lemniscate

  The lemniscate is a curve that looks like either a figure 8 or an infinity sign, depending on its orientation. The equation for the horizontal (infinity sign) version is $r^2 = a\cos(2\theta)$, and the equation for the vertical version is $r^2 = a\sin(2\theta)$. Figure 3.9 is a diagram of the horizontal version, for your viewing pleasure!

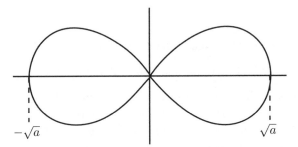

Figure 3.9 $r^2 = a\cos(2\theta)$

- The Archimedean Spiral

  It's not surprising that making spirals in polar coordinates isn't too difficult. The equation $r = a + b\theta$ does the job the way that Archimedes was particularly fond of, as shown in figure 3.10.

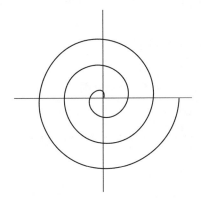

Figure 3.10 $r = a + b\theta$

Varying the a varies the orientation of the spiral, and varying the b varies how "tightly" the spiral is wound.

If you're lucky enough to have access to a graphical calculator, please don't fight the urge to go and play with the parameters on all of these curves, and see what shapes you can make!

EXERCISES

3.3.1. Express the Cartesian point $(1, 1, 1)$ in cylindrical polar coordinates.

3.3.2. Express the Cartesian point $(0, 2, 2)$ in spherical polar coordinates.

3.3.3. What is the equation of a vertical semicircle of radius 1, centred at the origin, and whose projection onto the x, y-plane lies on the Cartesian line $y = x$ in the first quadrant? Give your answer in spherical polar coordinates.

3.3.4. What is the equation of a horizontal circle of radius 4 and a constant height of 3, whose projection onto the x, y-plane is centred at the origin? Give your answer in cylindrical polar coordinates.

3.3.5. What is the equation of a horizontal circle of radius 3 and a constant height of -3, whose projection onto the x, y-plane is centred at the origin? Give your answer in spherical polar coordinates.

3.3.6. What is the general polar equation of the shape that looks like the infinity sign?

3.3.7. What is the general polar equation of the shape that looks like a figure 8?

3.3.8. What is the general polar equation of a limaçon?

3.3.9. What is the general polar equation of a cardioid?

3.3.10. What type of shape is defined by an equation of the form $r = a + b\theta$?

Where Now?

In this chapter we've dealt mainly with how points and shapes are *expressed* using polar coordinates. I'm sure you'll agree that simply being able to *express* a shape in Cartesian coordinates is far from being an expert in their use, and so naturally there is a lot more to look at with polar coordinates than we have covered here. Using changes of coordinate systems is a vital tool in many calculus-based problems.

In physics, one of the most useful applications of polar coordinates comes when dealing with surfaces, particularly in calculating something called flux. Flux is the rate that a given vector field "cuts through" a surface – you may well have heard the term when dealing with magnetism, where flux describes magnetic field density and the area that it is acting upon. When we mathematically calculate flux, it is necessary to parameterise (i.e., describe in a specific way) surfaces before we begin any calculations, and the best route to parameterising tricky surfaces often involves the use of cylindrical or spherical polar coordinates. You'll learn lots more about this, and many of the other uses of polar coordinates, in your university studies.

Vector Calculus (J. Marsden and A. Tromba, W.H. Freeman and Company, 2003) is an excellent introduction to the many ways in which polar coordinates (both in two and three dimensions) can help us solve calculus-based problems with shapes that aren't easily dealt with in the Cartesian system. If you had some trouble getting to grips with cylindrical or spherical polar coordinates, there's a great exploration of both in *Vector Calculus*, along with some interesting notes about the history of their development.

All that remains is, as promised, the proof of the polar coordinates formula for a straight line that doesn't pass through the origin. We'll be using Figure 3.11, where l is the line we're trying to describe, l_0 is the straight line through the origin which is also perpendicular to l, and l_1 is any straight line passing through the origin.

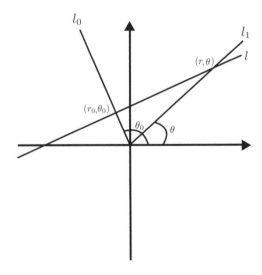

Figure 3.11

- Pick l_0: the line through the origin that is perpendicular to l. This means that l_0 has the equation $\theta = \theta_0$.

- Find the point at which l and l_0 intersect. This point is the point (r_0, θ_0).

- Note that any point on l can be expressed as (r, θ).

- l_1 is any line that passes through the origin, so the angle between l_1 and l is either $(\theta - \theta_0)$ or $(\theta_0 - \theta)$, depending on where we chose l_1 to be.

- Using standard trigonometry we arrive at either $r \cos(\theta - \theta_0) = r_0$ or $r \cos(\theta_0 - \theta) = r_0$ (again depending on where we chose l_1 to be).

- Because cos is an even function [i.e., $\cos(x) = \cos(-x)$], we have that $\cos(\theta - \theta_0) = \cos(\theta_0 - \theta)$. So it doesn't matter where we chose l_1 to be; we end up with the same equation.

- Rearranging what we got above yields $r = \frac{r_0}{\cos(\theta - \theta_0)}$.

That's all, folks!

$\mathit{4}$

Complex Numbers

Test Yourself

If you think you are already comfortable with this material, try these questions first and mark them using the answers at the back of the book. If you get them all right, you're probably ready to move straight on to the next chapter. If some look tricky, study the chapter first and then come back to these when you're ready.

1. What do each of $\mathbb{N}$, $\mathbb{Z}$, $\mathbb{Q}$, $\mathbb{R}$ and $\mathbb{C}$ mean?

2. Solve the equation $x^2 = -81$.

3. Solve the equation $x^2 - 6x = -10$.

4. Simplify $(7 + 3i) - (6 - 2i)$.

5. Simplify $(3 + 2i)(4 - 5i)$.

6. Simplify $\frac{6+2i}{3+5i}$.

7. Simplify $\frac{4+i}{-1-3i}$.

8. Draw $(7 + i)$ and $(-4 + 3i)$ on an Argand diagram (also known as "the complex plane").

9. Express $1 + \sqrt{3}i$ in $re^{i\theta}$ form.

10. By first expressing $(-\sqrt{3} - i)$ in $re^{i\theta}$ form, find $(-\sqrt{3} - i)^4$ in $a + bi$ form.

E. Hurst and M. Gould, *Bridging the Gap to University Mathematics*,
DOI: 10.1007/978-1-84800-290-6_4,
© Springer-Verlag London Limited 2009

4.1 Numbers

As a student in school, ending a solution of equations question by writing "no real roots" always seemed a bit of a disappointment. All of that work getting there, only to be stopped in our tracks by the problem that no real number, when squared, will yield a negative. But look again carefully at what we write: "No *real* roots." In the number system that we are using to answer the question, the real numbers, we can't find a solution. But what if there were some other number system – something new entirely – that allowed us to conquer this difficulty?

The most basic set of numbers is the natural numbers, $\mathbb{N}$. These are the positive, whole numbers $1, 2, 3, \ldots$ With these numbers, we can undertake the useful tasks of addition and multiplication: whenever we add or multiply a natural number with another, we get a natural number as a result.

But is addition and multiplication of whole, positive numbers enough? No, of course it isn't: let's look at subtraction. The simple subtraction $7 - 12$ does not have a "solution" if we restrict ourselves to natural numbers. To overcome this problem, we have a set of numbers called the integers, $\mathbb{Z}$. These are the positive *and* negative whole numbers, and 0. Working with the integers, we can be sure that we can undertake addition, multiplication and subtraction with confidence.

Of the four basic mathematical operations, lastly comes division. While the integers will deal with *exact division* such as $\frac{8}{8}$ or $\frac{-6}{3}$, anything with a remainder after division is not catered for by $\mathbb{Z}$. For this reason, we have the set of rational numbers, $\mathbb{Q}$ (the choice of the letter Q relates to the word "quotient," stating that all rational numbers can be expressed as a fraction). The rational numbers are very useful, and in everyday life the rational numbers is the set of numbers that serve the general public very well.

For those few of us that require a little bit more, we have the real numbers, $\mathbb{R}$. The reals are the rational numbers with some extra "snazzy bits" thrown in: things like π and e, and also things like $\sqrt{2}$. We need to use analysis, which you'll study at great length at university, in order to define the real numbers fully. There's an introduction to analysis in the last two chapters of this book.

And so, for a reasonable range of mathematics, the reals are *adequate*. They do everything we'd ever want to do, up to a point. Here's a quick diagram to illustrate the number systems we've looked at so far:

$$\mathbb{N} \xrightarrow{\text{Subtraction}} \mathbb{Z} \xrightarrow{\text{Division}} \mathbb{Q} \xrightarrow{\text{``Analysis''}} \mathbb{R}$$

But wait: there's more.

The Magic i

As we have seen, every "bigger" set of numbers is used to accommodate our need for more and more complicated calculations. So why stop at the reals? We know that solving $\sqrt{-1}$ causes problems in the reals, so again we're going to need a "bigger" set of numbers: the complex numbers, $\mathbb{C}$.

Simply by defining $\sqrt{-1} = i$, we overcome most of the problems we have with reals. Now, what is $\sqrt{-16}$? Well, $\sqrt{-16} = \sqrt{16}\sqrt{-1} = 4i$. Notice, however, that if in the set of complex numbers we didn't include all the real numbers as well as i, we wouldn't be able to solve something like $\sqrt{4}$. For this reason, we have:

A complex number is of the form a+bi, where a and b are real numbers and $i = \sqrt{-1}$.

Have a good read of that and *remember it*. You'll be using that from day one at university, and you'll use it for the rest of your mathematical career. The a part of the number is called the "real" part, and the b part of the number is referred to as the "imaginary" part.

The $a + bi$ form is really helpful when we use our old friend, the quadratic formula:

$$\frac{-b \pm \sqrt{b^2 - 4ac}}{2a}$$

Note that we can write this as $\frac{-b}{2a} \pm \frac{\sqrt{b^2-4ac}}{2a}$, which will *always* yield a result in the form $a + bi$. The b might be 0, but it's still in the $a + bi$ form, so never again will we need to write "no real roots"!

Here are a couple of examples:

- Solve $x^2 + 25 = 0$

$$x^2 = -25$$
$$x = \pm\sqrt{-25}$$
$$= \pm\sqrt{25}\sqrt{-1}$$
$$= \pm 5i$$

- Solve $x^2 - 4x + 5 = 0$

$$\text{Use the quadratic formula: } x = \frac{4 \pm \sqrt{16 - 20}}{2}$$

$$= \frac{4}{2} \pm \frac{\sqrt{-4}}{2}$$

$$= 2 \pm \frac{\sqrt{4}\sqrt{-1}}{2}$$

$$= 2 \pm i$$

EXERCISES

4.1.1. What do α, β, γ and δ represent in this diagram?

$$\mathbb{N} \xrightarrow{\alpha} \mathbb{Z} \xrightarrow{\beta} \mathbb{Q} \xrightarrow{\gamma} \mathbb{R} \xrightarrow{\delta} \mathbb{C}$$

4.1.2. What is i^2?

4.1.3. Fill in the blanks:

> A complex number is of the form ... where ... and ... are ... numbers and ... $= \sqrt{-1}$.

4.1.4. Solve the equation $x^2 = -100$.

4.1.5. Solve the equation $x^2 + 64 = 0$.

4.1.6. Solve the equation $x^2 - 2x + 2 = 0$.

4.1.7. Solve the equation $x^2 + 4x + 20 = 0$.

4.1.8. Solve the equation $8x^2 - 4x + 1 = 0$.

4.1.9. Solve the equation $x^2 - 2x + 3 = 0$.

4.1.10. Solve the equation $3x^2 - 4x = -3$.

4.2 Working with Complex Numbers

Now that we've seen where complex numbers come from, and know that they're of the form $a + bi$, where a and b are real numbers and $i = \sqrt{-1}$, the next thing to look at is how we work with our new best friends.

Addition and Subtraction

You'll be delighted to know that addition and subtraction is *easy*. You see, $i = \sqrt{-1}$, and so we'll never affect the imaginary part of the number by adding or subtracting the real parts, and we'll never affect the real part of the number by adding or subtracting the imaginary parts (remember, when a complex number is written in the $a + bi$ form, a is the "real part" and b is the "imaginary part"). This means that when we add or subtract complex numbers, we just deal with the real part and the imaginary part *separately*, like this:

- Add $(3 + 6i)$ to $(5 + 9i)$

$$(3 + 6i) + (5 + 9i) = (3 + 5) + (6i + 9i)$$
$$= 8 + 15i$$

- If $x = (7 - 3i)$ and $y = (6 + 2i)$, find $x - y$.

$$(7 - 3i) - (6 + 2i) = (7 - 6) + (-3i - 2i)$$
$$= 1 - 5i$$

Very nice indeed, I'm sure you'll agree!

Multiplication

Multiplication really isn't too bad, so long as you take care. The key fact that we will be using is $i^2 = -1$, because $i = \sqrt{-1}$. So long as we keep that in mind, we shouldn't have too many problems.

- Simplify $(6 + i)(4 + 3i)$.

$$(6 + i)(4 + 3i) = 24 + 18i + 4i + 3i^2$$
$$= 24 + 22i + 3i^2$$
$$= 24 + 22i - 3$$
$$= 21 + 22i$$

- Simplify $(3 + i)(4 - 2i)$.

$$(3 + i)(4 - 2i) = 12 - 6i + 4i - 2i^2$$
$$= 12 - 2i - 2i^2$$
$$= 12 - 2i + 2$$
$$= 14 - 2i$$

Division

Division is the only time when we need to be a bit more cunning. The process is very similar to how we work with square roots in reals: the key is that *we don't want any funny business going on in the denominator*. If we could always get the denominator to be a real number, we'd be laughing. Lucky for us, there's a trick that lets us do exactly that.

A *complex conjugate* is a complex number in the $a + bi$ form, but with the sign of the imaginary part changed. That might not be too clear in words, so here are some examples:

Complex Number	Complex Conjugate
$a + bi$	$a - bi$
$a - bi$	$a + bi$
$-a + bi$	$-a - bi$
$-a - bi$	$-a + bi$

Make sure that you've noticed that the sign before the real part doesn't change, only the one on the imaginary part.

Let's take a look at what happens if we multiply a complex number by its complex conjugate:

$$(a + bi)(a - bi) = a^2 - abi + abi - b^2 i^2$$
$$= a^2 + b^2$$

Do you see why this is useful? There's no i term anymore. Now let's see how we can apply this in the division of complex numbers:

- Find $\frac{(3+2i)}{(5-3i)}$.

To try to do this division blindly (without our neat trick) is futile. We're going to need complex conjugates to take this one down, so let's see how we're going to proceed.

Remember our motto: *we don't want any funny business going on in the denominator*. So let's use complex conjugates to knock that denominator into shape.

You might have guessed this by now: we're going to multiply top and bottom of the fraction by the complex conjugate of the denominator:

$$\frac{(3 + 2i)}{(5 - 3i)} = \frac{(3 + 2i)(5 + 3i)}{(5 - 3i)(5 + 3i)}$$

Now, simplifying each of the numerator and the denominator independently:

$$\frac{(3+2i)(5+3i)}{(5-3i)(5+3i)} = \frac{15+9i+10i+6i^2}{25+15i-15i-9i^2}$$
$$= \frac{15+19i-6}{25+9}$$
$$= \frac{9+19i}{34}$$

At this point, we're done! The number is in the complex form: it is $\frac{9}{34} + \frac{19}{34}i$. With some careful work and the awesome trick, we have the answer.

One final example here before a jumbo set of exercises. You'll definitely need these skills early on in your degree, so make sure that you work through them all!

- Find $\frac{(4-6i)}{(-5+i)}$.

$$\frac{(4-6i)}{(-5+i)} = \frac{(4-6i)(-5-i)}{(-5+i)(-5-i)}$$
$$= \frac{-20-4i+30i+6i^2}{25+5i-5i-i^2}$$
$$= \frac{-26+26i}{26} = -1+i$$

EXERCISES

4.2.1. Simplify $(6+3i)+(4+5i)$.

4.2.2. Simplify $(7-3i)+(2i)$.

4.2.3. Simplify $(3-2i)-(3-2i)$.

4.2.4. Simplify $(4+i)-(7-11i)$.

4.2.5. Simplify $(12+3i)+(3-12i)$.

4.2.6. Simplify $(6+3i)(6+3i)$.

4.2.7. Simplify $(7-i)(3+2i)$.

4.2.8. Simplify $(18+3i)(1-i)$.

4.2.9. Simplify $(12-i)(3+2i)$.

4.2.10. Simplify $(6+2i)(3+i)(4-i)$.

4.2.11. Simplify $\frac{6+2i}{2}$.

4.2.12. Simplify $\frac{3+2i}{1+i}$.

4.2.13. Simplify $\frac{3+5i}{2-2i}$.

4.2.14. Simplify $\frac{10-i}{7+3i}$.

4.2.15. Simplify $\frac{-3-i}{-2+i}$.

4.2.16. Simplify $\left(\frac{-2+i}{2-i}\right)\left(\frac{4+i}{-1+i}\right)$.

4.3 Tips and Tricks

As well as the standard a + bi form of a complex number, there are a few "tricks" that we can use in order to cut out some of the lengthy calculations that we occasionally have to do.

Argand Diagrams

We need to acquaint ourselves with Argand diagrams (sometimes known, less imaginatively, as "the complex plane"). By now, you'll be very familiar with the standard (x, y) plane. This plane is useful to us, because it lets us plot a piece of data with two parts as a single point in the space. Well, our new buddies complex numbers are each a single piece of data with two parts (the real part and the imaginary part), so why not plot them in a similar way?

Notice first that to make later work clearer, we're not going to plot complex numbers as a single dot. Instead, we're going to draw a straight line from the origin out to that point. Here's an example:

Plot $(3 + 2i)$, $(-4 + 3i)$ and $(-1 - 4i)$ on an Argand diagram.

Figure 4.1 shows how this is done. Also notice which way around the axes are labelled: The real part is on the x axis, the imaginary part is on the y-axis.

The $re^{i\theta}$ Form

In order to follow this next bit, you're going to need to be up to speed with the basic idea of polar coordinates. If you missed that chapter, it's pretty much necessary to go back there and work through it. Doing so will make what follows much more logical for you.

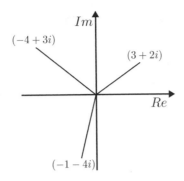

Figure 4.1

Recall that as well as defining a point in the plane by its (x, y)-co-ordinates, we can also uniquely determine it by its distance from the origin and the angle that it makes with the positive x-axis. Now, for a point plotted on an Argand diagram, expressing it in this polar form (called "the $re^{i\theta}$ form") is very helpful indeed. Here's an example:

Plot $2 + 2i$ on an Argand diagram. Express $2 + 2i$ in $re^{i\theta}$ form, and hence find $(2 + 2i)^6$.

Obviously, multiplying out $(2+2i)(2+2i)(2+2i)(2+2i)(2+2i)(2+2i)$ (like we would need to if it weren't for the trick we're about to explore) is not too much fun. Figure 4.2 shows the plot the question asked for.

To go into polar coordinates, we need to find how far this point is from the

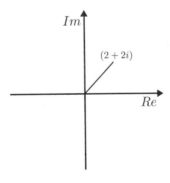

Figure 4.2

origin and the angle it makes with the positive x-axis.

To get the distance, let's give Pythagoras a call: take a glance at Figure 4.3.

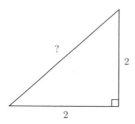

Figure 4.3

Pythagoras seems pretty confident that $2\sqrt{2}$ is the distance from the origin.

To get the angle from the positive x-axis (labelled "real" in an Argand diagram), trigonometry seems as good a weapon as any: $\tan\alpha = \frac{2}{2}$, and hence $\alpha = \frac{\pi}{4}$. So we can express our point as being $2\sqrt{2}$ away from the origin, and $\frac{\pi}{4}$ from the positive x-axis.

At this point, it's necessary to have to ask for your trust: we're going to *state* something without proof. The reasons are twofold: firstly, the trick we're learning is just that – a handy tool to reduce tedious calculation. We aren't doing this for the theory, so learning to *do* it is what we're after, not learning where it's from. The second reason is much more important: When your university lecturer proves *why* we can do what we're about to do, they'll no doubt take a quick detour to a special case: $e^{\pi i}$. Some people find what happens in this special case unremarkable, but to many – ourselves included – the $e^{\pi i}$ moment is something like an artist seeing the *Mona Lisa* for the first time: a mind blowing, perhaps life-changing experience where everything you've ever learned suddenly seems worthwhile. Anyway, enough sentiment, on with the maths. Like we say, trust us on this one:

> We make the $re^{i\theta}$ form by setting our distance as the r and our angle as the θ.

This means that after we've found the distance and angle, we're actually done. So the $2 + 2i$ example becomes $2\sqrt{2}e^{i\frac{\pi}{4}}$.

Now, upon gut reaction, this seems far nastier than simply "$2+2i$." But look at what we have here: $2\sqrt{2}e^{i\frac{\pi}{4}}$ *uniquely* determines our $2+2i$. Now look over the last part of the question: "and hence find $(2 + 2i)^6$." Our $2\sqrt{2}e^{i\frac{\pi}{4}}$ is *equivalent* to $2+2i$, so rather than writing $(2+2i)(2+2i)(2+2i)(2+2i)(2+2i)(2+2i)$, we can write $(2\sqrt{2}e^{i\frac{\pi}{4}})^6$, which is simply $(2\sqrt{2})^6(e^{i\frac{6\pi}{4}})$, which is $512e^{i\frac{3\pi}{2}}$. Now, if we need our answer in the $a + bi$ form again, we're just looking for a point on an Argand diagram that is 512 away from the origin, and at an angle of $\frac{3\pi}{2}$ from the positive x-axis. Visualising where this would lie on an Argand diagram, we see that our answer is $(0 - 512i)$, or simply $-512i$. I'm sure you'll agree that this method is a lot less painful than a massive binomial onslaught!

One more worked example, then your exercises await:

Use the $re^{i\theta}$ form of complex numbers to find $(\sqrt{3} + i)^{12}$.

Let's first find the length of the line:

$$\sqrt{(\sqrt{3})^2 + 1^2} = 2$$

Now, let's find the angle from the positive x-axis:

$$\tan \alpha = \frac{1}{\sqrt{3}}$$
$$\alpha = \frac{\pi}{6}$$

So our $re^{i\theta}$ form is $2e^{\frac{\pi}{6}i}$. Hence we require $(2e^{\frac{\pi}{6}i})^{12} = 4096e^{2\pi i}$.

We know an angle of 2π means our point lies exactly on the positive x-axis, and so $(\sqrt{3} + i)^{12} = 4096$ (there is no imaginary part to our answer).

EXERCISES

4.3.1. Draw an Argand diagram, and represent these points on it:
$\alpha = (7 + i), \beta = (3 - i), \gamma = (-1 + 3i), \delta = (-3 - 3i)$.

4.3.2. How far is $(3 + 2i)$ from the origin when drawn on an Argand diagram?

4.3.3. How far is $(-2 - i)$ from the origin when drawn on an Argand diagram?

4.3.4. What angle does $2i$ make with the positive x-axis when drawn on an Argand diagram?

4.3.5. What angle does $(-3 - 3i)$ make with the positive x-axis when drawn on an Argand diagram?

4.3.6. Express $(-2 + 2i)$ in $re^{i\theta}$ form.

4.3.7. Express $(-\sqrt{3} + i)$ in $re^{i\theta}$ form.

4.3.8. By first expressing $(-3 + 3i)$ in $re^{i\theta}$ form, find $(-3 + 3i)^4$ in $a + bi$ form.

Where Now?

Complex numbers are a *vital* tool in all sorts of mathematics, and throughout your career as a mathematician you'll see them *all over the place*. You'll be using them to solve differential equations; you'll be using them to find eigenvalues (explored later in this book!); later in your degree you'll be studying their properties as a ring... the list goes on.

If you're interested in complex numbers and how they work, this is definitely an area that you can brush up on before going to university. In this chapter we worked with the $re^{i\theta}$ form, but there's also the very useful $r(\cos\theta + i\sin\theta)$ form too: you can find out where this comes from and how to use it in *A Concise Introduction to Pure Mathematics* (M. Liebeck, Chapman and Hall/CRC, 2000).

If you like the idea of complex numbers in the abstract sense (rather than simply what we can use them for), you may also be interested in quaternions. In calculations they work in a similar way to complex numbers, but they have a few more complications to consider: *The Foundations of Mathematics* (I. Stewart and D. Tall, Oxford University Press, 1977) is a great book to delve into if you'd like to know a little more about this topic (but be warned, working with quaternions gets quite difficult quite quickly!).

5
Vectors

Test Yourself

If you think you are already comfortable with this material, try these questions first and mark them using the answers at the back of the book. If you get them all right, you're probably ready to move straight on to the next chapter. If some look tricky, study the chapter first and then come back to these when you're ready.

1. Find the norm of the vector $(3, 8, 3, 3, 3)$.

2. Given that the vector $\mathbf{u} = (\frac{1}{6}, \frac{1}{4}, x, \frac{1}{3})$ is a unit vector, find the value of x.

3. If $\mathbf{m} = (3, 7, 5)$ and $\mathbf{n} = (8, 1, 3)$, find $3\mathbf{m} - 2\mathbf{n}$.

4. If $\mathbf{y} = (3, 7, 1, 3)$, find $\|\mathbf{y}\|\mathbf{y}$.

5. Let $\mathbf{u} = 8\mathbf{i} + 2\mathbf{j} + 6\mathbf{k}$ and $\mathbf{v} = 9\mathbf{i} + 5\mathbf{j} + \mathbf{k}$. Find $2\mathbf{u} + 3\mathbf{v}$.

6. Find $(1, 6, 3, 8) \cdot (4, 2, 6, 1)$.

7. Let $\mathbf{a} = (1, 4, 3)$, $\mathbf{b} = (2, 4, 1)$ and $\mathbf{c} = (7, 1, 0)$. Find $\mathbf{a}(\mathbf{b} \cdot \mathbf{c})$.

8. Find the angle between the vectors $(2, 2, 1)$ and $(2, 3, 6)$. Leave your answer in a trigonometric form.

9. Find the cross-product of the vectors $(2, 3, 4)$ and $(1, 3, 5)$.

10. Find $((2, 2, 7) \times (3, 7, 8)) \cdot (5, 6, 7)$.

E. Hurst and M. Gould, *Bridging the Gap to University Mathematics*,
DOI: 10.1007/978-1-84800-290-6_5,
© Springer-Verlag London Limited 2009

5.1 Reinventing the Wheel

From a reasonably early stage in "serious" mathematics, vectors play a central role. In this chapter, we're going to look at lots of different uses of vectors, and learn some tricks that we can employ when we are working with them. They should be a relatively familiar concept, but first let's quickly review some of the crucial tools in our conquest to solve vector problems.

Vectors and Euclidean n-space

You may have met vectors in $\mathbb{R}^2$ before – that is to say, vectors in the form (a, b), where a and b are both real numbers. A vector like this can be thought of as the line, with direction, from the origin to the point P, where $P = (a, b)$. Therefore, this vector can be written as $\vec{OP}$, as well as the standard form of (a, b). The same idea can be employed for vectors in $\mathbb{R}^3$ (3D space).

For vectors in $\mathbb{R}^n$, where $n > 3$, it is not possible to think visually (don't try, it will give you headaches), but these vectors can be written and manipulated in the same way as vectors in $\mathbb{R}^2$ and $\mathbb{R}^3$. Here's a quick definition that will make life easier when we proceed:

> The set $\mathbb{R}^n$ (that is the set of vectors of the form $(x_1, x_2, x_3, \ldots, x_n)$, where $x_1, x_2, x_3, \ldots, x_n$ are all real numbers) is called Euclidean n-space.

Unsurprisingly, Euclid was the first person to widely use the notion of a vector in this way!

Bits and Pieces

A few standard definitions:

- A vector $\mathbf{u} = (u_1, u_2)$ is equal to another vector $\mathbf{v} = (v_1, v_2)$ if *and only if* $u_1 = v_1$ and $u_2 = v_2$: that is, their components are equal.

- The zero vector in $\mathbb{R}^2$ is simply $\mathbf{0} = (0, 0)$: the vector containing 2 0s. Note that $\mathbf{0}$ *is not* equal to the number 0.

- The length of a vector in $\mathbb{R}^2$ is found using Pythagoras' theorem, so for a vector $\mathbf{v} = (v_1, v_2)$, the length is $\sqrt{v_1^2 + v_2^2}$.

Now, most of your previous work on vectors was probably only in $\mathbb{R}^2$ or $\mathbb{R}^3$, but can we extend these definitions for use on all vectors in Euclidean n-space?

Let's try to adapt what we already have for $\mathbb{R}^2$.

In $\mathbb{R}^2$, we said that we declare vectors to be equal if their two components are equal. To extend this idea to Euclidean n-space is straightforward – we'll just demand that two vectors can only be equal if all of their components are equal. So the vector $(3, 7, 9, 6)$ is *only* equal to the vector $(3, 7, 9, 6)$, and it would be impossible for the vector $(7, 8, 3, 2)$ to be equal to a vector with five components, because *all* of the components cannot be equal.

In $\mathbb{R}^2$, the $\mathbf{0}$ vector is $(0, 0)$. In $\mathbb{R}^3$ it is $(0, 0, 0)$. It makes good logical sense, then, to define the $\mathbf{0}$ vector in Euclidean n-space as the vector with n 0s.

The only real difficulty with all of these definitions comes with length. In $\mathbb{R}^2$, we can certainly find the "length" of any vector by Pythagoras – we can extend this idea to $\mathbb{R}^3$ too. But what about beyond that? We already know that we can't "visualise" $\mathbb{R}^4$ because the world that we know is 3-dimensional. Suddenly, the notion of "length" is no good to us.

Despite this, finding the root of the sum of the squares (quite a mouthful!) is still useful to us in $\mathbb{R}^4$ and beyond, so we have a different name for it: the Euclidean norm – we'll just write "norm" for short, but there are plenty of other kinds of norm out there that you'll meet during your degree studies. This norm is denoted by double vertical lines either side of the vector, and is defined as:

$$||\mathbf{v}|| = \sqrt{v_1^2 + v_2^2 + v_3^2 + \cdots + v_n^2}.$$

From now on, even in $\mathbb{R}^2$ and $\mathbb{R}^3$, we're going to call this quantity "norm," not "length." Here's a quick example:

Find the norm of the vector $\mathbf{v} = (2, 5, 6, 2, 10)$.

$$\begin{aligned}
||\mathbf{v}|| &= \sqrt{2^2 + 5^2 + 6^2 + 2^2 + 10^2} \\
&= \sqrt{4 + 25 + 36 + 4 + 100} \\
&= \sqrt{169} \\
&= 13
\end{aligned}$$

Here is a helpful idea that we can get from computing the norm:

A unit vector is a vector that has a norm equal to 1.

So $\mathbf{v}$ is a unit vector if *and only if* $\sqrt{v_1^2 + v_2^2 + v_3^2 + \cdots + v_n^2} = 1$. Unit vectors turn out to be helpful in all kinds of sticky situations in degree mathematics, so it's a good idea to learn this definition now!

Vector Addition and Scalar Multiplication

When dealing with vectors, we often also need to deal with ordinary numbers at the same time. We use the word *scalar* to mean "a number." We're going to take a look at vector addition and scalar multiplication, but these processes should already be reasonably familiar in $\mathbb{R}^2$ and $\mathbb{R}^3$, so let's jump straight to working in Euclidean n-space.

For vectors $\mathbf{u} = (u_1, u_2, u_3, \ldots, u_n)$ and $\mathbf{v} = (v_1, v_2, v_3, \ldots, v_n)$, and some scalar λ, we define:

- $\mathbf{u} + \mathbf{v} = (u_1 + v_1, u_2 + v_2, u_3 + v_3, \ldots, u_n + v_n)$.

- $\mathbf{u} - \mathbf{v} = (u_1 - v_1, u_2 - v_2, u_3 - v_3, \ldots, u_n - v_n)$.

- $\lambda\mathbf{u} = (\lambda u_1, \lambda u_2, \lambda u_3, \ldots, \lambda u_n)$.

So, for example, if $\mathbf{u} = (1, 4, 2, 8, 7)$ and $\mathbf{v} = (8, 3, 6, 2, 7)$, then:

$$
\begin{aligned}
\mathbf{u} + \mathbf{v} &= (1 + 8, 4 + 3, 2 + 6, 8 + 2, 7 + 7) \\
&= (9, 7, 8, 10, 14) \\
\mathbf{u} - \mathbf{v} &= (1 - 8, 4 - 3, 2 - 6, 8 - 2, 7 - 7) \\
&= (-7, 1, -4, 6, 0) \\
4\mathbf{u} &= (4, 16, 8, 32, 28)
\end{aligned}
$$

Notice that vector addition (and subtraction) is only defined for when we have two vectors with the same number of components – otherwise, there is nothing that we can do with them.

It is sometimes helpful to think about vectors in $\mathbb{R}^2$ geometrically (i.e., in terms of pictures). By doing this, we can see the ideas of vector addition and scalar multiplication at work. Figure 5.1 shows two vectors, $\mathbf{u}$ and $\mathbf{v}$, and Figure 5.2 shows the result of undertaking $\mathbf{u} + \mathbf{v}$.

In vector addition, we can see that the final vector can be found by "completing the parallelogram": the sum of the two vectors is simply the diagonal of the parallelogram that they form when placed end to end.

Vector subtraction can also be performed geometrically: Figure 5.3 shows the result of $\mathbf{u} - \mathbf{v}$. In vector subtraction, if we imagine "reversing" the direction of the vector that we're subtracting, we again find that the result of the subtraction can be seen by placing the vectors end to end, and finding the vector from the origin to the endpoint.

Figure 5.4 shows result of $2\mathbf{v}$. Here we find the result by simply drawing the same vector, end to end, twice!

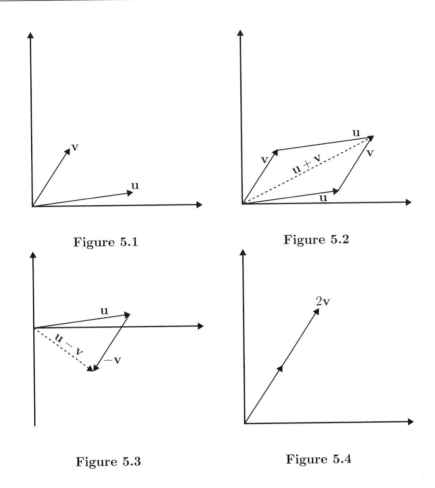

Figure 5.1   Figure 5.2

Figure 5.3   Figure 5.4

Important Properties

In most of the mathematics that we do, we take certain things for granted. Before university we are often careless about stating what things we are assuming when we make calculations, and this perhaps misguides students into wrongly believing that they aren't assuming anything at all. Think about the pair of real numbers, m and n. We often use the fact $m + n = n + m$, but it actually turns out to be something that we call an axiom: we use this fact all over the place, but we can't actually *prove* it – instead, we have to *define* that our numbers behave in this way. This is exactly what an axiom is: something that we *insist* a certain set of numbers does, because we need the result to be true. In the case of $m + n = n + m$, we'll never find a counterexample because we have *defined* that the numbers we use behave in this manner.

There is a set of axioms that we need when we work with vectors. Remember them well, because if you don't have them listed as part of the syllabus when you start your course on linear algebra, you probably missed a lecture. Sleep at night next time – you have better dreams that way.

Let $\mathbf{u}$, $\mathbf{v}$ and $\mathbf{w}$ be vectors in $\mathbb{R}^n$, and λ and μ be scalars. Then the axioms that we need are as follows:

1. $\mathbf{u} + \mathbf{v} = \mathbf{v} + \mathbf{u}$.

2. $(\mathbf{u} + \mathbf{v}) + \mathbf{w} = \mathbf{u} + (\mathbf{v} + \mathbf{w})$.

3. $\mathbf{u} + \mathbf{0} = \mathbf{u}$.

4. $\mathbf{u} + (-\mathbf{u}) = \mathbf{0}$.

5. $(\lambda\mu)\mathbf{u} = \lambda(\mu\mathbf{u})$.

6. $(\lambda + \mu)\mathbf{u} = \lambda\mathbf{u} + \mu\mathbf{u}$.

7. $\lambda(\mathbf{u} + \mathbf{v}) = \lambda\mathbf{u} + \lambda\mathbf{v}$.

8. $1\mathbf{u} = \mathbf{u}$.

9. $(-1)\mathbf{u} = -\mathbf{u}$.

10. $0\mathbf{u} = \mathbf{0}$.

Remember to keep an eye out for the difference between 0 and $\mathbf{0}$.

EXERCISES

5.1.1. Find the norm of the vector $(-5, 6, -2, 9, -5, 5)$.

5.1.2. What value of x makes $(\frac{2}{13}, \frac{5}{13}, x, \frac{2}{13}, \frac{10}{13})$ a unit vector?

5.1.3. Find $(3, 6, 2, 4) + (-2, 8, 3, -9)$.

5.1.4. Find $(5, -2, -8, 3, 8) - (4, -3, 5, -7, -5)$.

5.1.5. Find $5(4, 7, 3, 8)$.

5.1.6. Find $3(2, 5, 6) - (8, 2, 9)$.

5.1.7. Find $2(-5, 8, 4, 2) + 8(-8, 4, 6, -3)$.

5.1.8. Find $10(5, 7, 2) - 3(-2, 5, 2) + (7, 3, 9)$.

5.1.9. Find $\|(6, 2, -9)\|(4, 2, 7)$.

5.1.10. Let $\mathbf{u} = (2, 4, -4, 6)$ and $\mathbf{v} = (2, -1, 3, -2)$. Find $\|\mathbf{u}\|\mathbf{u} - \|\mathbf{v}\|\mathbf{v}$.

5.2 A Different Approach

Sometimes it is helpful to think of vectors in a more algebraic form. All vectors in $\mathbb{R}^2$ can be expressed in terms of the two vectors $\mathbf{i}$ and $\mathbf{j}$, where $\mathbf{i} = (1, 0)$ and $\mathbf{j} = (0, 1)$. The vector $\mathbf{v} = (a, b)$ can be simply written $\mathbf{v} = a\mathbf{i} + b\mathbf{j}$. A specific example of this might be a question that asks you to write the vector $\mathbf{u} = (6, 2)$ in the $\mathbf{i}, \mathbf{j}$ form. The answer to this would be $\mathbf{u} = 6\mathbf{i} + 2\mathbf{j}$. Not too tough, is it?

Similarly, vectors in $\mathbb{R}^3$ can be written in terms of the $\mathbf{i}, \mathbf{j}$ and $\mathbf{k}$ vectors, where $\mathbf{i} = (1, 0, 0)$, $\mathbf{j} = (0, 1, 0)$ and $\mathbf{k} = (0, 0, 1)$. So in $\mathbb{R}^3$, the vector $\mathbf{m} = (a, b, c)$ can be written as $\mathbf{m} = a\mathbf{i} + b\mathbf{j} + c\mathbf{k}$, and the vector $\mathbf{n} = (1, 8, 5)$ can be written as $\mathbf{n} = \mathbf{i} + 8\mathbf{j} + 5\mathbf{k}$.

Having vectors in this form makes addition and scalar multiplication no harder. For example, if $\mathbf{a} = 8\mathbf{i} + 2\mathbf{j} + 9\mathbf{k}$ and $\mathbf{b} = 3\mathbf{i} - 9\mathbf{j} + 3\mathbf{k}$, then we can work like this:

$$\mathbf{a} + \mathbf{b} = (8 + 3)\mathbf{i} + (2 - 9)\mathbf{j} + (9 + 3)\mathbf{k}$$
$$= 11\mathbf{i} - 7\mathbf{j} + 12\mathbf{k}$$
$$3\mathbf{a} = (3 \times 8)\mathbf{i} + (3 \times 2)\mathbf{j} + (3 \times 9)\mathbf{k}$$
$$= 24\mathbf{i} + 6\mathbf{j} + 27\mathbf{k}$$

For those of you that prefer to think in algebraic forms, this might seem a bit more intuitive than how we looked at vectors in the previous section, as now we simply collect the coefficients in $\mathbf{i}, \mathbf{j}$ and $\mathbf{k}$ separately, just as if we were working with a standard equation in three unknowns.

Totally Dotty

Although many people encounter the dot product before university, we're going to approach the idea in Euclidean n-space. Here's the definition:

> If $\mathbf{u} = (u_1, u_2, \ldots, u_n)$ and $\mathbf{v} = (v_1, v_2, \ldots, v_n)$, then $\mathbf{u} \cdot \mathbf{v} = u_1 v_1 + u_2 v_2 + \cdots + u_n v_n$.

If you haven't met the dot product before, make sure you take extra care with that definition. Notice that what we get from it is *not* a vector: we're summing lots of scalars, so we're going to get just a scalar as the result. If you prefer to think of this sort of process in words, then we are *multiplying corresponding components, then summing the results*. Here's an example:

> Find $\mathbf{u} \cdot \mathbf{v}$, where $\mathbf{u} = (2, 7, 5, 9, 1)$ and $\mathbf{v} = (7, 2, 6, 4, 4)$.

$$\mathbf{u} \cdot \mathbf{v} = (2 \times 7) + (7 \times 2) + (5 \times 6) + (9 \times 4) + (1 \times 4)$$
$$= 14 + 14 + 30 + 36 + 4$$
$$= 98$$

Properties of Dotting

Here are five helpful properties to bear in mind when working with the dot product, which hold true for all vectors $\mathbf{u}, \mathbf{v}$ and $\mathbf{w}$ and scalars λ:

1. $\mathbf{u} \cdot \mathbf{v} = \mathbf{v} \cdot \mathbf{u}$.

2. $(\lambda\mathbf{u}) \cdot \mathbf{v} = \lambda(\mathbf{u} \cdot \mathbf{v}) = \mathbf{u} \cdot (\lambda\mathbf{v})$.

3. $\mathbf{u} \cdot (\mathbf{v} + \mathbf{w}) = \mathbf{u} \cdot \mathbf{v} + \mathbf{u} \cdot \mathbf{w}$.

4. $\mathbf{u} \cdot \mathbf{0} = 0$.

5. $\mathbf{u} \cdot \mathbf{u} = ||\mathbf{u}||^2$.

Angles Between Vectors

There is a very useful identity that we can use to find the angle between two vectors in Euclidean n-space. This rule is crucial, so remember it well:

$$\mathbf{u} \cdot \mathbf{v} = ||\mathbf{u}||||\mathbf{v}|| \cos\theta$$

So long as we know what $\mathbf{u}$ and $\mathbf{v}$ are, the only unknown in this identity is θ, so we can definitely find it. Here's an example:

Find the angle between the vectors $\mathbf{u} = (0, 3, 0)$ and $\mathbf{v} = (1, 1, \sqrt{2})$.

We know the rule is $\mathbf{u} \cdot \mathbf{v} = ||\mathbf{u}||||\mathbf{v}|| \cos\theta$, so rearranging we get:

$$\cos\theta = \frac{\mathbf{u} \cdot \mathbf{v}}{||\mathbf{u}||||\mathbf{v}||}$$

We first need to find the values of $\mathbf{u} \cdot \mathbf{v}$, $||\mathbf{u}||$ and $||\mathbf{v}||$:

$$\mathbf{u} \cdot \mathbf{v} = (0 \times 1) + (3 \times 1) + (0 \times \sqrt{2})$$
$$= 0 + 3 + 0$$
$$= 3$$
$$||\mathbf{u}|| = \sqrt{0^2 + 3^2 + 0^2}$$
$$= \sqrt{9}$$
$$= 3$$
$$||\mathbf{v}|| = \sqrt{1^2 + 1^2 + \sqrt{2}^2}$$
$$= \sqrt{4}$$
$$= 2$$

This means that:

$$\cos\theta = \frac{3}{6}$$
$$= \frac{1}{2}$$
$$\theta = \frac{\pi}{3}$$

There's one final definition to round this section off:

Two vectors are said to be *orthogonal* if the angle between them is $\frac{\pi}{2}$.

Now note that if $\theta = \frac{\pi}{2}$, then $\cos\theta = 0$, so for any two vectors $\mathbf{u}$ and $\mathbf{v}$ to be orthogonal, we *must* have $\mathbf{u}\cdot\mathbf{v} = 0$. Thinking about the dot product, this is quite helpful in that if we have two nonzero vectors whose dot product is 0, we know *for certain* that these vectors must be orthogonal, because $\mathbf{u}\cdot\mathbf{v} = ||\mathbf{u}||||\mathbf{v}||\cos\theta$, and $||\mathbf{u}||$ and $||\mathbf{v}||$ will *never* equal 0 when $\mathbf{u}$ and $\mathbf{v}$ are nonzero vectors, so $\cos\theta$ *must* equal 0. When you're happy with the logic here, you're ready to tackle the next batch of exercises:

EXERCISES

5.2.1. Write $(2, 7, 3)$ in the $\mathbf{i}, \mathbf{j}, \mathbf{k}$ notation.

5.2.2. If $\mathbf{u} = 8\mathbf{i} - 2\mathbf{j} - 6\mathbf{k}$ and $\mathbf{v} = 3\mathbf{i} - 9\mathbf{j} + 4\mathbf{k}$, find $\mathbf{u} + \mathbf{v}$ and $5\mathbf{u}$.

5.2.3. Find $\mathbf{u} \cdot \mathbf{v}$, where $\mathbf{u} = (4, 2, 7, 2, 6, 4)$ and $\mathbf{v} = (9, 0, 2, 2, 1, 8)$.

5.2.4. Find $\mathbf{u} \cdot \mathbf{v}$, where $\mathbf{u} = (8, 2, 6)$ and $\mathbf{v} = (9, 2, 5)$.

5.2.5. Let $\mathbf{u} = (2, 7, 4)$, $\mathbf{v} = (8, 2, 1)$ and $\mathbf{w} = (1, 5, 4)$. Find $\mathbf{u} \cdot (\mathbf{v} + \mathbf{w})$.

5.2.6. Let $\mathbf{u} = (2, 1, 7)$, $\mathbf{v} = (7, 2, 9)$ and $\mathbf{w} = (2, 1, 4)$. Find $(\mathbf{u} \cdot \mathbf{v})\mathbf{w}$.

5.2.7. Find the angle between the vectors $(3, 4)$ and $(5, 12)$. Leave your answer in a trigonometric form.

5.2.8. Find the angle between the vectors $(6, 8)$ and $(8, 15)$. Leave your answer in a trigonometric form.

5.3 The Cross Product

Imagine now that we have two vectors, and we need to find a third vector that is orthogonal to *both* of these (remember, in three dimensions this is perfectly sane!). Rather than having to try to visualise things and fumble around in the dark, we have a sneaky tool.

The cross product is one of those rare things in mathematics that is not universally applicable. You see, there is a strict constraint on it: we can only use the cross product when we have two vectors in $\mathbb{R}^3$, and we wish to find a third vector that is orthogonal to both. As such, it's quite a limited tool, but it is still very handy – plus there is the added bonus that when we come on to working with matrices, we can actually employ the same computational procedure and get a *very* useful result. The notation for "the cross product" is simply a "times" sign, so $\mathbf{a} \times \mathbf{b}$ is "the cross product of $\mathbf{a}$ and $\mathbf{b}$." Performing $\mathbf{a} \times \mathbf{b}$ (or, indeed, $\mathbf{b} \times \mathbf{a}$) will give us a vector orthogonal to both $\mathbf{a}$ and $\mathbf{b}$.

Anyway, here's how to do it: because it's simply a tool to be used, don't worry about where it comes from, just learn the steps and be able to apply them.

1. Take your two vectors in $\mathbb{R}^3$ and write them out *one beneath the other.* So if we were finding the cross product $\mathbf{u} \times \mathbf{v}$, where $\mathbf{u} = (a, b, c)$ and $\mathbf{v} = (d, e, f)$, we'd start by writing:

$$(a, b, c)$$
$$(d, e, f)$$

Note that it's important that the first vector in the cross product is written on top.

2. Now, we're going to view what we have as three separate columns: $\begin{pmatrix} a \\ d \end{pmatrix}$, $\begin{pmatrix} b \\ e \end{pmatrix}$ and $\begin{pmatrix} c \\ f \end{pmatrix}$.

3. Imagine "ignoring" the first column (i.e. $\begin{pmatrix} a \\ d \end{pmatrix}$). We're left with $\begin{pmatrix} b \\ e \end{pmatrix}$ and $\begin{pmatrix} c \\ f \end{pmatrix}$.

4. With the remaining columns $\begin{pmatrix} b \\ e \end{pmatrix}$ and $\begin{pmatrix} c \\ f \end{pmatrix}$, we do the calculation "top of first times bottom of second minus top of second times bottom of first." In our algebraic example, this means computing $bf - ce$. This is the first component of our solution vector.

5. Now, to get the second component, we follow the same idea with only a minor change: Ignore the *middle* column instead of the first, and of what's remaining do "top of first times bottom of second minus top of second times bottom of first." Also, with this second component, we want *minus* the result we get from this calculation. In our algebraic example, this gives us $-(af - cd)$.

6. Finally, we're going to take the positive result of our calculation when ignoring the *third* column as the third component of the solution vector. This is $ae - bd$.

7. We now have our solution vector: $(bf - ce, -(af - cd), ae - bd)$.

If you don't remember the process and just memorise this result then you're not really losing out on too much. Have a couple of reads through, make sure that you know what the result should be, and then head off into the final set of exercises.

EXERCISES

5.3.1. Find the cross product of $(3, 2, 5)$ and $(2, 7, 5)$.

5.3.2. Find a vector that is orthogonal to both $(2, 1, 5)$ and $(3, 2, 1)$.

5.3.3. Find a vector that is orthogonal to both $(1, 4, 6)$ and $(2, 2, 5)$.

5.3.4. Find a vector that is orthogonal to both $(4, 6, 2)$ and $(5, 1, 3)$.

Where Now?

The study of vectors leads very naturally into the study of matrices, which it's quite possible that you'll never have met before. You'll be much better prepared to tackle matrices if you're very confident about what's going on with vectors. *Vector Calculus* (J. Marsden and A. Tromba, W.H. Freeman and Company,

2003) is a great source of more examples to practice on if you feel that you want to develop your skills a little further.

Also, vectors are very important in their own right. They allow us to look at lots of pieces of data as a single vector, and so can greatly cut down our calculation work. In addition to all of this, there are lots of little "hidden benefits" to some of the skills developed in this chapter. Just one such example is that if we express two adjacent sides of a parallelogram in vector form, then the norm of the cross product of these two vectors is equal to the area of the parallelogram!

By the time you arrive at university, you'll no doubt be a master of differentiating all sorts of real functions. When you start your degree, there's a great deal of work to be done in order to *prove* that all of the tools you're using are indeed valid. In doing this work, you'll uncover a difficulty: differentiating gets problematic when there are vectors involved. Many "old favourites" like the chain rule need a complete overhaul to work with vectors – and that's just the tip of the iceberg. Once again, *Vector Calculus* helps develop many of the skills needed to tackle calculus problems with vectors.

6
Matrices

Test Yourself

If you think you are already comfortable with this material, try these questions first and mark them using the answers at the back of the book. If you get them all right, you're probably ready to move straight on to the next chapter. If some look tricky, study the chapter first and then come back to these when you're ready.

1. Find $\begin{pmatrix} 2 & 3 \\ 9 & 2 \end{pmatrix} + \begin{pmatrix} 5 & 6 \\ -3 & -2 \end{pmatrix}$.

2. Find $3\begin{pmatrix} 8 & -2 \\ 9 & 5 \\ 4 & 6 \end{pmatrix} - \begin{pmatrix} -2 & 4 \\ 3 & -1 \\ 8 & 1 \end{pmatrix}$.

3. Find $4\begin{pmatrix} 2 & 1 \\ 3 & -1 \end{pmatrix} + 6\begin{pmatrix} -1 & 6 \\ -4 & 2 \end{pmatrix}$.

4. Find $\begin{pmatrix} 2 & 3 \\ -1 & 4 \end{pmatrix}\begin{pmatrix} 5 & -6 \\ 1 & 2 \end{pmatrix}$.

5. Find $\begin{pmatrix} 2 & -1 \\ -3 & 4 \\ -2 & 9 \end{pmatrix}\begin{pmatrix} 2 & -6 \\ 4 & 1 \end{pmatrix}$.

E. Hurst and M. Gould, *Bridging the Gap to University Mathematics*,
DOI: 10.1007/978-1-84800-290-6_6,
© Springer-Verlag London Limited 2009

6. Find $\begin{pmatrix} 9 & -5 & 3 \\ -2 & 1 & 6 \\ 5 & 4 & -1 \end{pmatrix} \begin{pmatrix} 2 & 4 \\ -3 & 1 \\ 6 & 2 \end{pmatrix} + \begin{pmatrix} 2 & 6 \\ -4 & 1 \\ 5 & 9 \end{pmatrix}$.

7. Find $\begin{pmatrix} -2 & 1 \\ 9 & 3 \end{pmatrix} \begin{pmatrix} 4 & -1 & 9 \\ -3 & 2 & 3 \end{pmatrix} + \begin{pmatrix} -6 & 3 & 1 \\ 1 & 5 & 7 \end{pmatrix}$.

8. Let $A = \begin{pmatrix} 3 & 4 \\ -2 & 1 \end{pmatrix}$. Find $\det(A)$.

9. Find the inverse of $\begin{pmatrix} 4 & 6 \\ -5 & -8 \end{pmatrix}$.

10. Let $A = \begin{pmatrix} 6 & 19 \\ 3 & 9 \end{pmatrix}$. Find A^{-1}.

6.1 Enter the Matrix

Unless you've already studied a lot of pure mathematics, it is quite unlikely that you'll have come across matrices before. They are actually one of the most important and useful tools in a whole range of mathematics, and as such they will appear in lots of your modules at university. This means, of course, that it is vital to get a really good understanding of the concepts and applications of matrices, because you will be able to use them in a whole range of situations – and therefore get lots of marks in exams!

A What?

If you've never met one before, you'll no doubt be wondering what a matrix actually *is* (other than a computerised world designed by the machines). Here's the definition:

A matrix is a rectangular array of numbers.

Not too awful, is it? (The definition, that is, not the joke). Here are some examples:

$$A = \begin{pmatrix} 1 & 5 \\ 3 & 4 \end{pmatrix}, B = \begin{pmatrix} 7 & 11 & 3 \\ 4 & 8 & 7 \end{pmatrix}, C = \begin{pmatrix} 8 & 12 & 9 \\ 9 & 6 & 3 \\ 1 & 5 & 7 \end{pmatrix}$$

If a matrix has m rows and n columns, then we call that matrix an $m \times n$ matrix. So, in the examples above, A is a 2×2 matrix, B is a 2×3 matrix, and C is a 3×3 matrix. This "$m \times n$" term is called the "order" of the matrix and, unsurprisingly, if $m = n$ then we call the matrix a "square matrix."

When we are talking about matrices in general, we can write out something like $A = \begin{pmatrix} \alpha_{11} & \alpha_{12} \\ \alpha_{21} & \alpha_{22} \end{pmatrix}$. In this case it is clear that we are talking about a 2×2 matrix, but we can abbreviate this and still convey the same information. We can instead write $A = (\alpha_{ij})_{m \times n}$, which says that we have a matrix of order $m \times n$ in which the intersection of the ith row and the jth column is the entry α_{ij}. Make sure that you're comfortable with the idea that both of these representations of a "general" matrix convey the same information – one in a visual way, the other in a concise way. Which one of these is used is largely based on the preference of its scribe, but sometimes one or the other of them is better at conveying a particular idea or argument.

Manipulation

Just like numbers and vectors, we can do lots of different things with a pair of matrices. Here, we're going to go through some of the basic operations that I know you're just *dying* to learn about. Firstly, let's tackle equality: rather like vectors, two matrices are equal if *and only if* their components are equal. So if we know that:

$$\begin{pmatrix} a & b \\ c & d \end{pmatrix} = \begin{pmatrix} e & f \\ g & h \end{pmatrix}$$

then we know that $a = e$, $b = f$, $c = g$ and $d = h$. Notice that by this definition two matrices of different orders can *never* be equal.

Now let's consider matrix addition. To add two matrices together, we simply add the two corresponding elements of each matrix. In algebraic language, this means that given a matrix $A = (\alpha_{ij})_{m \times n}$ and $B = (\beta_{ij})_{m \times n}$, then the (i, j)th component of the matrix $A + B$ is $\alpha_{ij} + \beta_{ij}$. Don't worry if you're not entirely happy with that statement right away; there are a lot of letters flying around. Take a glance at this worked example, then come back for a second look:

$$\text{If } A = \begin{pmatrix} 5 & 3 \\ 1 & 4 \end{pmatrix} \text{ and } B = \begin{pmatrix} 2 & 1 \\ 1 & 8 \end{pmatrix}, \text{ find } A + B.$$

$$A + B = \begin{pmatrix} 5 & 3 \\ 1 & 4 \end{pmatrix} + \begin{pmatrix} 2 & 1 \\ 1 & 8 \end{pmatrix}$$

$$= \begin{pmatrix} 5+2 & 3+1 \\ 1+1 & 4+8 \end{pmatrix}$$

$$= \begin{pmatrix} 7 & 4 \\ 2 & 12 \end{pmatrix}$$

Matrix subtraction is performed in the same way, except that we *subtract* the corresponding elements of the matrices. The logic these mathematicians use – I just don't understand it sometimes! Here's the worked example:

If $C = \begin{pmatrix} 8 & 0 \\ 6 & 9 \end{pmatrix}$ and $D = \begin{pmatrix} 2 & 1 \\ 5 & 3 \end{pmatrix}$, find $C - D$.

$$\begin{pmatrix} 8 & 0 \\ 6 & 9 \end{pmatrix} - \begin{pmatrix} 2 & 1 \\ 5 & 3 \end{pmatrix} = \begin{pmatrix} 8-2 & 0-1 \\ 6-5 & 9-3 \end{pmatrix}$$

$$= \begin{pmatrix} 6 & -1 \\ 1 & 6 \end{pmatrix}$$

As you might well have noticed, there is a strict constraint on matrix addition and subtraction: *We can only add or subtract matrices of the same order.* There are no exceptions to this, so if you have two matrices of differing orders, then it *just ain't happening.*

Scalar Multiplication

Another very straightforward manipulation with matrices is that of *scalar multiplication*. If you've studied the chapter on vectors you'll recall that a scalar is "just a number," and so scalar multiplication of matrices is anything of the form λA, where λ is the scalar and A is the matrix. To perform scalar multiplication of matrices, we simply *multiply every element in the matrix by the scalar.* Here's a quick example to illustrate the idea:

$$3 \begin{pmatrix} 2 & 4 \\ 1 & 9 \end{pmatrix} = \begin{pmatrix} (3 \times 2) & (3 \times 4) \\ (3 \times 1) & (3 \times 9) \end{pmatrix} = \begin{pmatrix} 6 & 12 \\ 3 & 27 \end{pmatrix}$$

And here's the same principle being applied to a 3×3 matrix:

$$-2 \begin{pmatrix} 4 & 9 & 6 \\ 1 & 3 & -5 \\ 0 & -2 & 14 \end{pmatrix} = \begin{pmatrix} (-2 \times 4) & (-2 \times 9) & (-2 \times 6) \\ (-2 \times 1) & (-2 \times 3) & (-2 \times -5) \\ (-2 \times 0) & (-2 \times -2) & (-2 \times 14) \end{pmatrix}$$

$$= \begin{pmatrix} -8 & -18 & -12 \\ -2 & -6 & 10 \\ 0 & 4 & -28 \end{pmatrix}$$

That really is all there is to it!

Before we go on to the more complicated operations, let's enjoy what should be a relatively straightforward batch of exercises.

EXERCISES

6.1.1. $A = \begin{pmatrix} \alpha_{11} & \alpha_{12} & \alpha_{13} & \alpha_{14} & \alpha_{15} \\ \alpha_{21} & \alpha_{22} & \alpha_{23} & \alpha_{24} & \alpha_{25} \\ \alpha_{31} & \alpha_{32} & \alpha_{33} & \alpha_{34} & \alpha_{35} \end{pmatrix}$ is one way of writing "A is a 3×5 matrix." What is the other way?

6.1.2. What is the order of A in the question above?

6.1.3. Find $\begin{pmatrix} 2 & 3 \\ 1 & 5 \end{pmatrix} + \begin{pmatrix} 5 & 6 \\ 2 & 1 \end{pmatrix}$.

6.1.4. Find $\begin{pmatrix} -8 & 1 \\ 7 & 4 \end{pmatrix} + \begin{pmatrix} -2 & 6 \\ -4 & 1 \end{pmatrix}$.

6.1.5. Find $3 \begin{pmatrix} 8 & 1 & 4 \\ -2 & 6 & 0 \\ 5 & 1 & -2 \end{pmatrix}$.

6.1.6. Find $\begin{pmatrix} 4 & 12 & -2 \\ 6 & 1 & 0 \\ 4 & 12 & 7 \end{pmatrix} - \begin{pmatrix} 11 & 3 & 6 \\ -4 & 1 & 12 \\ 3 & -3 & 1 \end{pmatrix}$.

6.1.7. If $B = \begin{pmatrix} 1 & 1 & -7 \\ 3 & 71 & 1 \\ 6 & -2 & 17 \end{pmatrix}$ and $C = \begin{pmatrix} 2 & 5 & 7 \\ 12 & 15 & 10 \end{pmatrix}$, is it possible to find either of $B + C$ or $B - C$? Why?

6.1.8. Find $8 \begin{pmatrix} -1 & 3 \\ 1 & 4 \end{pmatrix} + \begin{pmatrix} 9 & 0 \\ 2 & 4 \end{pmatrix}$.

6.1.9. Find $6 \begin{pmatrix} 1 & 3 & 2 \\ -1 & 4 & 9 \end{pmatrix} - 3 \begin{pmatrix} 1 & 7 & 4 \\ 4 & 2 & 1 \end{pmatrix}$.

6.1.10. Find $2 \begin{pmatrix} 1 & 8 & -4 \\ -6 & -9 & 1 \\ 2 & 5 & 6 \end{pmatrix} + \begin{pmatrix} 4 & -3 & 6 \\ 2 & 5 & -9 \\ -1 & 4 & 3 \end{pmatrix}$.

6.2 Multiplication and More

Now that we've looked at matrix addition and matrix subtraction, the next logical step is to examine matrix multiplication. Sadly, this one is not as easy as addition or subtraction, but don't despair because once you have the hang of it, it's not too bad at all.

Again, let's start with the algebraic definition first and then take a look at the process in action, so that it's clearer what the definition really *means*:

If $A = (\alpha_{ij})_{l \times m}$ and $B = (\beta_{ij})_{m \times n}$, then we define AB to be the matrix $C = (\gamma_{ij})_{l \times n}$, such that $\gamma_{ij} = \alpha_{i1}\beta_{1j} + \alpha_{i2}\beta_{2j} + \cdots + \alpha_{im}\beta_{mj}$.

Unless you have an excellent eye for algebra that will no doubt look confusing, so let's look at it carefully using a worked example.

$$\text{Let } A = \begin{pmatrix} 2 & 0 & 5 \\ 1 & 3 & 8 \end{pmatrix} \text{ and } B = \begin{pmatrix} 1 & 3 & 0 \\ 5 & 2 & 1 \\ 6 & 4 & 0 \end{pmatrix}. \text{ Find } AB.$$

Let's go about breaking that definition down, bit by bit. From the algebraic definition, we can see that we *are* allowed to multiply these matrices: if we multiply a 2×3 matrix by a 3×3 matrix, then we will get a 2×3 matrix as our answer. If we call C the solution matrix (i.e., the matrix AB), then $C = \begin{pmatrix} \gamma_{11} & \gamma_{12} & \gamma_{13} \\ \gamma_{21} & \gamma_{22} & \gamma_{23} \end{pmatrix}$. Now, let's take a look at what we're going to actually have to *do* to find this solution matrix.

The "mess" of alphas, betas and gammas in the definition is simply showing us how we find specific elements in the solution matrix. Let's try to "decode" what's there, to find γ_{11}. We see that $\gamma_{11} = \alpha_{11}\beta_{11} + \alpha_{12}\beta_{21} + \alpha_{13}\beta_{31}$. So, in this case, $\gamma_{11} = (2 \times 1) + (0 \times 5) + (5 \times 6)$. That is, $\gamma_{11} = 32$.

The key thing to notice here is that *this process is the same as something that you've seen before*. So long as you've studied the chapter on vectors, you should recognise this as taking the *dot product* of the first row of A with the first column of B. Nice, eh? That definitely clears things up for us:

To find the (i, j)th element of the solution matrix, we simply take the dot product of the ith row of A with the jth column of B.

Let's quickly work through a few more, and find the solution matrix:

$$\begin{aligned} \gamma_{11} &= (2, 0, 5) \cdot (1, 5, 6) \\ &= (2 \times 1) + (0 \times 5) + (5 \times 6) \\ &= 32 \end{aligned}$$

$$\gamma_{12} = (2, 0, 5) \cdot (3, 2, 4)$$
$$= (2 \times 3) + (0 \times 2) + (5 \times 4)$$
$$= 26$$

$$\gamma_{13} = (2, 0, 5) \cdot (0, 1, 0)$$
$$= (2 \times 0) + (0 \times 1) + (5 \times 0)$$
$$= 0$$

$$\gamma_{21} = (1, 3, 8) \cdot (1, 5, 6)$$
$$= (1 \times 1) + (3 \times 5) + (8 \times 6)$$
$$= 64$$

$$\gamma_{22} = (1, 3, 8) \cdot (3, 2, 4)$$
$$= (1 \times 3) + (3 \times 2) + (8 \times 4)$$
$$= 41$$

$$\gamma_{23} = (1, 3, 8) \cdot (0, 1, 0)$$
$$= (1 \times 0) + (3 \times 1) + (8 \times 0)$$
$$= 3$$

The solution matrix to the worked example is $\begin{pmatrix} 32 & 26 & 0 \\ 64 & 41 & 3 \end{pmatrix}$. That might seem like an enormous effort, but as you get more and more practice at doing it, you won't need to write out all of the individual steps – doing it in your head and just writing down the answer is much faster. Before we take a look at another example, here is a list of some important things to bear in mind when conducting matrix multiplication:

- Although the two matrices don't have to be of the same order, we can only multiply matrices if the number of columns in the first matrix is the same as the number of rows in the second.

- The solution matrix will have the same number of rows as the first matrix, and the same number of columns as the second.

- From the two facts above, it should be clear that, in general, $AB \neq BA$. This is a key fact, where matrices behave differently to numbers. Don't forget it!

Here's one last worked example before we move on:

$$\text{Find} \begin{pmatrix} 1 & -2 & 5 \\ -2 & 1 & 3 \\ 0 & 3 & -1 \end{pmatrix} \begin{pmatrix} 2 & 5 & 0 \\ 3 & -1 & -3 \\ 6 & -1 & 0 \end{pmatrix}.$$

$$\gamma_{11} = (1, -2, 5) \cdot (2, 3, 6)$$
$$= 26$$

$$\gamma_{12} = (1, -2, 5) \cdot (5, -1, -1)$$
$$= 2$$

$$\gamma_{13} = (1, -2, 5) \cdot (0, -3, 0)$$
$$= 6$$

$$\gamma_{21} = (-2, 1, 3) \cdot (2, 3, 6)$$
$$= 17$$

$$\gamma_{22} = (-2, 1, 3) \cdot (5, -1, -1)$$
$$= -14$$

$$\gamma_{23} = (-2, 1, 3) \cdot (0, -3, 0)$$
$$= -3$$

$$\gamma_{31} = (0, 3, -1) \cdot (2, 3, 6)$$
$$= 3$$

$$\gamma_{32} = (0, 3, -1) \cdot (5, -1, -1)$$
$$= -2$$

$$\gamma_{33} = (0, 3, -1) \cdot (0, -3, 0)$$
$$= -9$$

So, the solution matrix is $\begin{pmatrix} 26 & 2 & 6 \\ 17 & -14 & -3 \\ 3 & -2 & -9 \end{pmatrix}$.

Diagonal Matrices

At the very beginning of this chapter we met square matrices: matrices with the same number of rows as columns. There is a special type of square matrix, called a *diagonal matrix*, which is a square matrix where all of the entries *not* on the "main diagonal" (i.e., any α_{ij} where $i \neq j$) are 0s. So that you can see what this means visually, here are some examples of diagonal matrices:

$$\begin{pmatrix} 2 & 0 \\ 0 & 5 \end{pmatrix}, \begin{pmatrix} 3 & 0 & 0 \\ 0 & 2 & 0 \\ 0 & 0 & 4 \end{pmatrix}, \begin{pmatrix} -7 & 0 & 0 & 0 \\ 0 & 1 & 0 & 0 \\ 0 & 0 & 0 & 0 \\ 0 & 0 & 0 & 14 \end{pmatrix}$$

If that last one (with a 0 in $(3,3)$) came as a surprise to you, take another look at the definition. We don't mind if 0s *are* on the diagonal, we only demand that there are 0s *everywhere not on the diagonal*. There is also an abbreviation, "*diag*," that is commonly used when referring to diagonal matrices (mathematicians can get viciously lazy when writing things down). We can refer to a diagonal matrix simply by the entries on its main diagonal, because we know that everywhere else is made up of 0s. So the matrices above can be written as diag$(2, 5)$, diag$(3, 2, 4)$ and diag$(-7, 1, 0, 14)$, respectively.

Identity Crisis

There is a special kind of diagonal matrix that is very useful when we work with matrices. It is called the *identity matrix*. When we multiply numbers together, the number 1 is "the multiplicative identity," because multiplying any number by 1 will give you the same number back. Unsurprisingly, then, the identity matrix is the matrix that you can multiply any other matrix by, and get that original matrix back again. For this reason, the identity matrix is a diagonal matrix with all 1s on the main diagonal. We write the identity matrix as I_n, where n is the number of rows (or columns – the matrix is square) in the matrix. Here are two examples of the identity matrix:

$$I_2 = \begin{pmatrix} 1 & 0 \\ 0 & 1 \end{pmatrix}, I_3 = \begin{pmatrix} 1 & 0 & 0 \\ 0 & 1 & 0 \\ 0 & 0 & 1 \end{pmatrix}$$

Let's check that the useful property of matrix multiplication by the identity matrix really does work:

$$\begin{pmatrix} 3 & 5 \\ 4 & 2 \end{pmatrix} \begin{pmatrix} 1 & 0 \\ 0 & 1 \end{pmatrix} = \begin{pmatrix} (3 \times 1) + (5 \times 0) & (3 \times 0) + (5 \times 1) \\ (4 \times 1) + (2 \times 0) & (4 \times 0) + (2 \times 1) \end{pmatrix} = \begin{pmatrix} 3 & 5 \\ 4 & 2 \end{pmatrix}$$

Interestingly, multiplying by the identity matrix is one of the few times where the order that multiplication is done in doesn't matter. You'll remember from the list above that, in general, $AB \neq BA$, but actually $AI_n = I_n A$. You can easily check this property by redoing the example above in the opposite order if you don't believe it!

Zilch

There's one last kind of "special" matrix that we're going to take a look at before we hit the exercises. This is the *zero matrix*, 0_n. It is an $(n \times n)$ matrix in which every element is 0. Hopefully these examples won't shock you:

$$0_2 = \begin{pmatrix} 0 & 0 \\ 0 & 0 \end{pmatrix}, 0_3 = \begin{pmatrix} 0 & 0 & 0 \\ 0 & 0 & 0 \\ 0 & 0 & 0 \end{pmatrix}$$

When we're dealing with numbers, 0 is "the additive identity," because adding 0 to any number will give you the same number back. The same is true of the zero matrix in matrix addition: adding the zero matrix to any other matrix will give the original matrix back. I'm sure that this is clear enough to not require an example, so let's get down to business.

EXERCISES

6.2.1. Find $\begin{pmatrix} 2 & 4 \\ 3 & 1 \end{pmatrix} \begin{pmatrix} 1 & 9 \\ 0 & 4 \end{pmatrix}$.

6.2.2. Find $\begin{pmatrix} 3 & 1 \\ 5 & 2 \end{pmatrix} \begin{pmatrix} 4 & 3 \\ 0 & 2 \end{pmatrix}$.

6.2.3. Find $\begin{pmatrix} 4 & 2 \\ 3 & 6 \\ 1 & 4 \end{pmatrix} \begin{pmatrix} 2 & 1 \\ 3 & 6 \end{pmatrix}$.

6.2.4. Find $\begin{pmatrix} -8 & 3 & 9 \\ 2 & 4 & -6 \end{pmatrix} \begin{pmatrix} 3 & -6 & 2 \\ 4 & 6 & -1 \\ 1 & -9 & 3 \end{pmatrix}$.

6.2.5. Find $\begin{pmatrix} 1 & 0 \\ 0 & 1 \end{pmatrix} \begin{pmatrix} 3 & 4 \\ 2 & 5 \end{pmatrix}$ and $\begin{pmatrix} 3 & 4 \\ 2 & 5 \end{pmatrix} \begin{pmatrix} 1 & 0 \\ 0 & 1 \end{pmatrix}$. What do you notice?

6.2.6. Find $\begin{pmatrix} 2 & 0 \\ 0 & 4 \end{pmatrix} \begin{pmatrix} -8 & 3 \\ 0 & 2 \end{pmatrix}$.

6.2.7. Find $\begin{pmatrix} -2 & 1 \\ 5 & 0 \end{pmatrix} \begin{pmatrix} 3 & 8 \\ 5 & 7 \end{pmatrix} + \begin{pmatrix} 2 & 4 \\ 1 & 3 \end{pmatrix}$.

6.2.8. Find $\begin{pmatrix} 1 & -6 & 5 \\ 0 & 1 & 2 \end{pmatrix} \begin{pmatrix} 1 & 0 \\ 8 & 4 \\ 9 & 0 \end{pmatrix} + \begin{pmatrix} -4 & 0 \\ 2 & -3 \end{pmatrix}$.

6.2.9. Find $\begin{pmatrix} 2 & 6 \\ 5 & 0 \end{pmatrix} \begin{pmatrix} 1 & 2 \\ 3 & 0 \end{pmatrix} + I_2$.

6.2.10. Find $\begin{pmatrix} 2 & 3 & 1 \\ 0 & 0 & 1 \\ 0 & 2 & 3 \end{pmatrix} \begin{pmatrix} 4 & 8 & 1 \\ 6 & 2 & 5 \\ 0 & 3 & 8 \end{pmatrix} - \text{diag}(1, 2, 3)$.

6.3 Determinants and Inverses

When dealing with square matrices, there is a very useful piece of information called *the determinant*. It has lots of different uses, and we're going to look at one of them now. In this whole section, we're going to restrict ourselves to 2×2 matrices, because finding the determinant of a square matrix bigger than this is much more complicated. First up, a definition:

The determinant of the matrix $\begin{pmatrix} a & b \\ c & d \end{pmatrix}$ is $ad - bc$.

There's not really anything to understand as such, so just remember it well. Much like the other things that we've just looked at, there are numerous different ways of writing "the determinant of matrix A." The most common is simply $\det(A)$, but if we want to write "the determinant of the matrix $A = \begin{pmatrix} a & b \\ c & d \end{pmatrix}$",

then we can also write $\begin{vmatrix} a & b \\ c & d \end{vmatrix}$ (that is, with straight lines at the edges, instead of the usual brackets). So, to quickly recap:

$$\det(A) = \begin{vmatrix} a & b \\ c & d \end{vmatrix} = ad - bc$$

Here are two worked examples of finding the determinant:

- $A = \begin{pmatrix} 2 & 5 \\ 3 & 4 \end{pmatrix}$. Find $\det(A)$.

$$\det(A) = (2 \times 4) - (5 \times 3) = 8 - 15 = -7.$$

- Find $\begin{vmatrix} 2 & 4 \\ 3 & 9 \end{vmatrix}$.

$$\begin{vmatrix} 2 & 4 \\ 3 & 9 \end{vmatrix} = (2 \times 9) - (4 \times 3) = 18 - 12 = 6.$$

Hopefully you've got that under your belt now, so let's look at a handy use for the determinant.

The Inverse of a Matrix

When dealing with real numbers, $\mathbb{R}$, every number (other than zero) has a "multiplicative inverse." That means for every real number we could ever imagine, there is another real number that we could multiply our number by, and get the multiplicative identity (i.e., 1) as the the result. In real numbers, finding the multiplicative inverse is really easy: If the number that we choose is x, then the multiplicative inverse is simply $\frac{1}{x}$ (which explains why we can't find a multiplicative inverse for zero). Now, let's consider a similar idea with matrices: if we take a matrix, might we be able to find another matrix so that when these two matrices are multiplied together we get the identity matrix?

As it turns out, the answer is *sometimes*. There are cases where a matrix has no inverse, but in the other cases (where an inverse *does* exist) there is a way to find it. There *is* a proof of what we're about to see, but understanding some of the steps requires a much greater knowledge of matrices than we've acquired here. Instead, we'll take a look at the result of the proof and then verify that this result does indeed have the required property: that a matrix multiplied by its inverse gives the identity matrix. Here's that result:

$$\text{If } A = \begin{pmatrix} a & b \\ c & d \end{pmatrix}, \text{ then the inverse of } A \text{ is } \tfrac{1}{\det(A)} \begin{pmatrix} d & -b \\ -c & a \end{pmatrix}.$$

In the statement above, we can clearly see when a matrix is invertible, and when it is not. For any 2×2 matrix, there is a $\frac{1}{\det(A)}$ scalar multiple. This means that if $ad = bc$, then $ad - bc = 0$, and hence $\frac{1}{\det(A)}$ is not defined. This is the factor that we use to work out whether a matrix is invertible or not: if $ad - bc = 0$, then the matrix is not invertible. If $ad - bc \neq 0$, then an inverse exists. Where it does exist, the shorthand for "the inverse of A" is simply A^{-1}.

Be sure to remember that this *doesn't* mean $\frac{1}{A}$, it means *the inverse of A*. Now let's check that all important property:

Verify that, for 2×2 matrices, $A^{-1}A = I_2$.

$$
\begin{aligned}
A^{-1}A &= \frac{1}{\det(A)} \begin{pmatrix} d & -b \\ -c & a \end{pmatrix} \begin{pmatrix} a & b \\ c & d \end{pmatrix} \\
&= \frac{1}{\det(A)} \begin{pmatrix} da - bc & bd - bd \\ -ac + ac & -bc + ad \end{pmatrix} \\
&= \frac{1}{\det(A)} \begin{pmatrix} da - bc & 0 \\ 0 & da - bc \end{pmatrix} \\
&= \frac{1}{\det(A)} \begin{pmatrix} \det(A) & 0 \\ 0 & \det(A) \end{pmatrix} \\
&= \begin{pmatrix} 1 & 0 \\ 0 & 1 \end{pmatrix}
\end{aligned}
$$

There we have it – a really neat little trick (remember that $ad - bc$ is the determinant of A), and we've verified what we need to. Of course, if we wanted to verify it with the A and A^{-1} the other way around (we've only done $A^{-1}A$ here, we should also do AA^{-1}), then exactly the same trick works – try it for yourself if you're confident. Here are a final few examples:

- Is the matrix $\begin{pmatrix} 4 & 6 \\ 2 & 3 \end{pmatrix}$ invertible? If it is, invert it.

    $ad - bc = (4 \times 3) - (6 \times 2) = 0$, so no, the matrix is not invertible.

- Is the matrix $\begin{pmatrix} -2 & -4 \\ 2 & 3 \end{pmatrix}$ invertible? If it is, invert it.

    $(-2 \times 3) - (-4 \times 2) = 2$. So $A^{-1} = \frac{1}{2} \begin{pmatrix} 3 & 4 \\ -2 & -2 \end{pmatrix} = \begin{pmatrix} \frac{3}{2} & 2 \\ -1 & -1 \end{pmatrix}$.

In the second case, where we *are* given an invertible matrix, we can check that our answer is correct by multiplying it by the original matrix, and checking that we get the identity matrix. Because we know that this should work whichever way around we do the multiplication, let's check it both ways:

$$
\begin{pmatrix} \frac{3}{2} & 2 \\ -1 & -1 \end{pmatrix} \begin{pmatrix} -2 & -4 \\ 2 & 3 \end{pmatrix} = \begin{pmatrix} -3 + 4 & -6 + 6 \\ 2 - 2 & 4 - 3 \end{pmatrix} = \begin{pmatrix} 1 & 0 \\ 0 & 1 \end{pmatrix}
$$

$$
\begin{pmatrix} -2 & -4 \\ 2 & 3 \end{pmatrix} \begin{pmatrix} \frac{3}{2} & 2 \\ -1 & -1 \end{pmatrix} = \begin{pmatrix} -3 + 4 & -4 + 4 \\ 3 - 3 & 4 - 3 \end{pmatrix} = \begin{pmatrix} 1 & 0 \\ 0 & 1 \end{pmatrix}
$$

All present and correct! Here's the final set of exercises:

EXERCISES

6.3.1. If $A = \begin{pmatrix} 2 & 3 \\ 4 & 1 \end{pmatrix}$, find the determinant of A.

6.3.2. If $B = \begin{pmatrix} -4 & 2 \\ 3 & 1 \end{pmatrix}$, find $\det(B)$.

6.3.3. Find $\begin{vmatrix} -2 & 3 \\ -1 & 6 \end{vmatrix}$.

6.3.4. Find $\begin{vmatrix} 4 & -9 \\ -6 & 2 \end{vmatrix}$.

6.3.5. Which of the following matrices are invertible?

$$A = \begin{pmatrix} 1 & 0 \\ 0 & 1 \end{pmatrix}, \quad B = \begin{pmatrix} 2 & -6 \\ -3 & -5 \end{pmatrix}, \quad C = \begin{pmatrix} -9 & 6 \\ 3 & -2 \end{pmatrix}$$

6.3.6. Find the inverse of $\begin{pmatrix} -3 & -13 \\ 6 & 10 \end{pmatrix}$.

6.3.7. Find the inverse of $\begin{pmatrix} -3 & 7 \\ 3 & -6 \end{pmatrix}$.

6.3.8. Let $A = \begin{pmatrix} -2 & 9 \\ -4 & -3 \end{pmatrix}$. Find A^{-1}.

Where Now?

Matrices have a massive range of uses, from differential equations to computer programming. We'll explore in the next chapter how matrices are fundamentally linked to linear maps, and another great use of them is in solving complicated systems of simultaneous equations. Much of the mathematical work that is done with computers relies heavily on matrices, and as such many areas of cutting edge research rely on them as a fundamental tool.

In finding determinants and inverses, we've only looked at the 2×2 case, but of course this is not the whole picture. The proof of *why* the inverse of a 2×2 matrix is what it is comes as part of the general case of invertible matrices.

It's definitely worth getting a good deal of practice with matrices before going to university. The early chapters of *Elementary Linear Algebra* (W. K. Nicholson, McGraw-Hill, 2004) have a good selection of exercises covering the things that we've done in this chapter, and the book also goes on to explore working with matrices in a more general sense.

The matrices that we looked at in this chapter all had finite order, but this doesn't have to be the case – from here, even the sky is not the limit.

7

Matrices as Maps

Test Yourself

If you think you are already comfortable with this material, try these questions
first and mark them using the answers at the back of the book. If you get them
all right, you're probably ready to move straight on to the next chapter. If some
look tricky, study the chapter first and then come back to these when you're
ready.

1. Find the matrix that corresponds to the linear map:

$$\begin{aligned} x_1 &= -2x_0 + y_0 \\ y_1 &= 4x_0 - 2y_0 \end{aligned}$$

2. Find the matrix that corresponds to the linear map:

$$\begin{aligned} x_1 &= -y_0 \\ y_1 &= 2x_0 + 3y_0 \end{aligned}$$

3. Write out the system of equations for the linear map whose corresponding
   matrix is $\begin{pmatrix} 3 & -2 \\ 0 & 1 \end{pmatrix}$.

4. What is the matrix corresponding to a reflection about the x-axis?

5. What is the matrix corresponding to a rotation by $\frac{\pi}{4}$ radians anticlockwise
   about the origin?

E. Hurst and M. Gould, *Bridging the Gap to University Mathematics*,
DOI: 10.1007/978-1-84800-290-6_7,
© Springer-Verlag London Limited 2009

6. Find the *single* matrix that corresponds to first rotating by π radians anti-clockwise about the origin, and *then* enlarging by a factor of 3 in both the horizontal and vertical directions.

7. $\begin{pmatrix} 3 \\ 4 \end{pmatrix}$ is an eigenvector of $\begin{pmatrix} 4 & 3 \\ 8 & 2 \end{pmatrix}$. What is the corresponding eigenvalue?

8. $\begin{pmatrix} 1 \\ -2 \end{pmatrix}$ is an eigenvector of $\begin{pmatrix} 4 & 3 \\ 8 & 2 \end{pmatrix}$. What is the corresponding eigenvalue?

9. Consider the matrix $A = \begin{pmatrix} 2 & 1 \\ 4 & -1 \end{pmatrix}$. What are its eigenvalues and corresponding eigenvectors?

10. Consider the matrix $B = \begin{pmatrix} -6 & 2 \\ -7 & 3 \end{pmatrix}$. What are its eigenvalues and corresponding eigenvectors?

7.1 Over and Over

In the previous chapter, we met a new beast: the matrix. So long as you've studied that chapter (and if you haven't, you should do so now!), you'll know the basics of performing matrix algebra. In this chapter we're going to take things a little further – we're going to look at the amazing relationship between matrices and *linear maps*. We're still going to stick with 2×2 matrices, but we're going to look at some of the very interesting things that we can do with them.

Every Journey Needs a Map

The idea of linear maps will almost certainly be new to you, as it's rarely covered by students before starting university. *Don't despair,* however; we're here to help.

Imagine that we have chosen a single point in the plane: let's call our point (x_0, y_0). We want to apply a map to our point, and doing so will "send" our point to a *new* point in the plane; we'll call our "destination point" (x_1, y_1). Imagine that we had a map so that the point we're mapping to is *uniquely* determined by the point at which we start, and furthermore that we could write the system of equations:

$$x_1 = ax_0 + by_0$$
$$y_1 = cx_0 + dy_0$$

where a, b, c and d are scalars – that is, we find (x_1, y_1) by taking some linear function of x_0 and y_0. Then the map that we're applying to (x_0, y_0) is called a *linear map*.

Exactly where the point (x_0, y_0) will be mapped *to* by applying the linear map is determined by the values of a, b, c and d that we choose. For this reason, keeping track of a, b, c and d is definitely worthwhile, but it would be helpful if we could somehow store them all together. We know two ways of doing this: we can either consider the vector (a, b, c, d), or we can consider a matrix with a, b, c and d as its entries. As it turns out, using matrices opens the doorway to a whole new way of thinking about problems.

The Matrix Multiverse

Let's put a, b, c and d into a matrix like this:

$$\begin{pmatrix} a & b \\ c & d \end{pmatrix}$$

We'll be going into more detail about many familiar linear maps later in this chapter, but let's take a look at one now. Imagine that we have chosen our point (x_0, y_0) and we want to use the linear map that maps our point *directly back onto itself*. That is, we want a map so that $(x_1, y_1) = (x_0, y_0)$: this is what we call the *identity map*. Remember our system of equations:

$$x_1 = ax_0 + by_0$$
$$y_1 = cx_0 + dy_0$$

We can see by equating coefficients (in this case by eye) that the solution we require to this system is when $x_1 = x_0$ and $y_1 = y_0$: that is when $a = 1$, $b = 0$, $c = 0$ and $d = 1$. Putting these into the 2×2 matrix yields:

$$\begin{pmatrix} 1 & 0 \\ 0 & 1 \end{pmatrix}$$

No prizes for spotting that this is *exactly* the matrix that we called "the identity matrix" in the previous chapter, but don't think of this as some kind of amazing coincidence. In actual fact, matrices are *defined* to act exactly like linear maps.

From Multiplication to Compositions

It would be helpful if we could rewrite the system of equations involving a, b, c and d as a single matrix equation, using the idea of matrix multiplication that we met in the previous chapter. Rather than trying to "work out" the equation, we'll state it and then prove that it does indeed represent the system that we require. Don't worry, this isn't cheating, it just speeds up the process a little: think of it like expanding brackets to prove that two expressions are equal, rather than factorising! The matrix equation that we're going to be working with is:

$$\begin{pmatrix} x_1 \\ y_1 \end{pmatrix} = \begin{pmatrix} a & b \\ c & d \end{pmatrix} \begin{pmatrix} x_0 \\ y_0 \end{pmatrix}$$

Multiplying out the right-hand side gives us:

$$\begin{pmatrix} x_1 \\ y_1 \end{pmatrix} = \begin{pmatrix} ax_0 + by_0 \\ cx_0 + dy_0 \end{pmatrix}$$

We saw in the last chapter that we declare two matrices to be equal if *and only if* all corresponding entries are equal, so for our system to have equality we need both of the following to hold:

$$x_1 = ax_0 + by_0$$
$$y_1 = cx_0 + dy_0$$

which is precisely the system of equations that we were describing earlier, so our matrix equation is definitely correct. If we set:

$$v_0 = \begin{pmatrix} x_0 \\ y_0 \end{pmatrix} \quad v_1 = \begin{pmatrix} x_1 \\ y_1 \end{pmatrix} \quad A = \begin{pmatrix} a & b \\ c & d \end{pmatrix}$$

then we can write our matrix equation as the much more elegant:

$$v_1 = Av_0$$

So we're comfortable dealing with a linear map which takes us from (x_0, y_0) to (x_1, y_1). Now let's imagine that we have *another* linear map, which takes us from (x_1, y_1) to (x_2, y_2). We can again write this as a matrix equation, like this:

$$\begin{pmatrix} x_2 \\ y_2 \end{pmatrix} = \begin{pmatrix} p & q \\ r & s \end{pmatrix} \begin{pmatrix} x_1 \\ y_1 \end{pmatrix}$$

Which we can further simplify by setting:

$$v_2 = \begin{pmatrix} x_2 \\ y_2 \end{pmatrix} \quad B = \begin{pmatrix} p & q \\ r & s \end{pmatrix}$$

Making our equation become:
$$v_2 = Bv_1$$

Now for the clever part – we can combine the equations:
$$v_1 = Av_0 \text{ and } v_2 = Bv_1$$

to get the single equation:
$$v_2 = BAv_0$$

Let's take a moment to step back and think about what's going on here: we considered a linear map taking us from (x_0, y_0) to (x_1, y_1), and then another that takes us from (x_1, y_1) to (x_2, y_2). But we could have considered the *single* linear map that took us from (x_0, y_0) *straight* to (x_2, y_2).

How would we write this as a matrix equation? Well, for some matrix C, it would be:
$$v_2 = Cv_0$$

But wait: that looks rather like our equation
$$v_2 = BAv_0$$

What's happening here?

Marvellous Matrices

The *composition* of the two linear maps results in us going straight from (x_0, y_0) to (x_2, y_2). Combining the matrix equations gave us the equation $v_2 = BAv_0$. Again, it's no coincidence that these two concepts yield strikingly similar results: *because composition of linear maps is equivalent to the multiplication of matrices.* That might seem a little strange, but to solidify things a little simply consider the matrix multiplication BA. If we let $BA = C$, then we arrive at the equation $v_2 = Cv_0$, which is precisely what we said we were looking for. Want things a little more formal than that? Well, here you go...

Examine the sets of equations that we had:
$$x_1 = ax_0 + by_0$$
$$y_1 = cx_0 + dy_0$$
$$x_2 = px_1 + qy_1$$
$$y_2 = rx_1 + sy_1$$

Plugging the first two equations into the second two gives us:
$$x_2 = p(ax_0 + by_0) + q(cx_0 + dy_0)$$
$$y_2 = r(ax_0 + by_0) + s(cx_0 + dy_0)$$

Expanding the brackets:

$$x_2 = pax_0 + qcx_0 + pby_0 + qdy_0$$
$$y_2 = rax_0 + scx_0 + rby_0 + sdy_0$$

Refactorising:

$$x_2 = (pa + qc)x_0 + (pb + qd)y_0$$
$$y_2 = (ra + sc)x_0 + (rb + sd)y_0$$

So our matrix BA is:

$$\begin{pmatrix} pa + qc & pb + qd \\ ra + sc & rb + sd \end{pmatrix}$$

This means that because we demand matrix multiplication to be equivalent to the composition of linear maps, the *only possible way* we could define matrix multiplication is:

$$\begin{pmatrix} p & q \\ r & s \end{pmatrix} \begin{pmatrix} a & b \\ c & d \end{pmatrix} = \begin{pmatrix} pa + qc & pb + qd \\ ra + sc & rb + sd \end{pmatrix}$$

EXERCISES

7.1.1. What name do we give to the linear map that is equivalent to the matrix $\begin{pmatrix} 1 & 0 \\ 0 & 1 \end{pmatrix}$?

7.1.2. Find the matrix that corresponds to the linear map:

$$\begin{aligned} x_1 &= - x_0 \\ y_1 &= y_0 \end{aligned}$$

7.1.3. Find the matrix that corresponds to the linear map:

$$\begin{aligned} x_1 &= 2x_0 + 3y_0 \\ y_1 &= x_0 - 4y_0 \end{aligned}$$

7.1.4. Find the matrix that corresponds to the linear map:

$$\begin{aligned} x_1 &= - 3x_0 \\ y_1 &= x_0 - 2y_0 \end{aligned}$$

7.1.5. Write out the system of equations for the linear map whose corresponding matrix is $\begin{pmatrix} 1 & 2 \\ 5 & 3 \end{pmatrix}$.

7.1.6. Write out the system of equations for the linear map whose corresponding matrix is $\begin{pmatrix} 2 & -3 \\ 6 & 2 \end{pmatrix}$.

7.1.7. Write out the system of equations for the linear map whose corresponding matrix is $\begin{pmatrix} 4 & 0 \\ 2 & 1 \end{pmatrix}$.

7.1.8. Where would the point $(7, 3)$ be mapped to by the linear map whose corresponding matrix is $\begin{pmatrix} 0 & 0 \\ 0 & 0 \end{pmatrix}$?

7.2 Old Friends

In this section, we'll be taking a closer look at some of the things that linear maps can do for us. You'll see that some of the transformations you've known for *years* can be achieved by matrix multiplication!

Mirror Images

Working on a 2-dimensional plane, one of the basic transformations that we might want to consider is reflection. Reflection is the process of taking the "mirror image" of the plane, about some given axis of reflection. We'll start by looking at reflection in lines which result in particularly nice matrices, and then we'll move onto a general case.

First up, we're going to look at reflection in the x-axis, as shown in Figure 7.1.

Let's consider exactly what a reflection like this does to our (x_0, y_0). The x coordinate is the same before and after the reflection, so we can immediately see that $x_1 = x_0$. The y coordinate, however, doesn't stay the same: its sign is reversed when the reflection takes place. This means that $y_1 = -y_0$. Putting these two equalities together gives us:

$$\begin{aligned} x_1 &= & x_0 \\ y_1 &= & - & y_0 \end{aligned}$$

This corresponds to the matrix:

$$\begin{pmatrix} 1 & 0 \\ 0 & -1 \end{pmatrix}$$

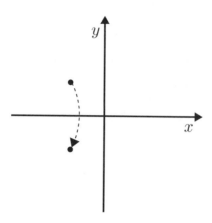

Figure 7.1

Similarly, we can reflect about the y-axis. This time our y coordinate is left unchanged by the reflection while the sign of the x coordinate changes, so our system of equations is:

$$
\begin{aligned}
x_1 &= - & x_0 \\
y_1 &= & y_0
\end{aligned}
$$

This corresponds to the matrix:

$$
\begin{pmatrix} -1 & 0 \\ 0 & 1 \end{pmatrix}
$$

Now that we've seen some specific cases of reflection, it's time to move towards a more general case. We're going to examine reflection about *any* line passing through the origin: the two lines above were certainly examples of that, so let's explore a little deeper. A good starting point would be refection about the line $y = x$, as in Figure 7.2.

Let's think carefully about what this reflection does to (x_0, y_0). After we apply the reflection, the x coordinate becomes the y coordinate and the y coordinate becomes the x coordinate. We therefore have the system of equations:

$$
\begin{aligned}
x_1 &= & y_0 \\
y_1 &= & x_0
\end{aligned}
$$

Which corresponds to the matrix:

$$
\begin{pmatrix} 0 & 1 \\ 1 & 0 \end{pmatrix}
$$

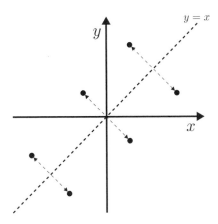

Figure 7.2

Excellent! We've now looked at a line that doesn't lie directly on one of the co-ordinate axes, but to finish the job entirely we need to find a way to describe a reflection in *any* straight line through the origin.

The key variable that we're going to need is the angle that the axis of reflection makes with the positive x-axis. We'll call this angle α, and measure it as shown in Figure 7.3.

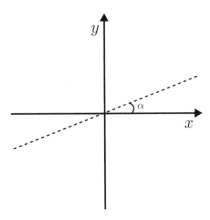

Figure 7.3

Let's derive the matrix that we require. We can imagine both our starting and end points as being the sum of an **x** vector (corresponding to the x component of that coordinate) and a **y** vector (corresponding to the y component of that coordinate), as in Figure 7.4.

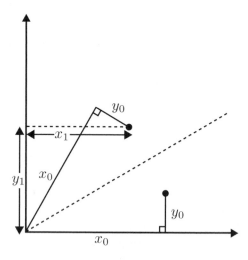

Figure 7.4

We've also drawn in the reflection of the $\mathbf{x}_0$ and $\mathbf{y}_0$ vectors, because we're going to call upon some trigonometry to express x_1 and y_1 in terms of x_0 and y_0, as in Figure 7.5.

In triangle PQR, we have:

$$\sin(2\alpha) = \frac{QR}{x_0}, \text{ and so } QR = x_0 \sin(2\alpha)$$

$$\cos(2\alpha) = \frac{PQ}{x_0}, \text{ and so } PQ = x_0 \cos(2\alpha)$$

For triangle RST:

$$P\hat{R}Q = \frac{\pi}{2} - 2\alpha$$

$$S\hat{R}T = 2\alpha \text{ because } P\hat{R}T = \frac{\pi}{2}$$

$$\sin(2\alpha) = \frac{ST}{y_0}, \text{ and so } ST = y_0 \sin(2\alpha)$$

$$\cos(2\alpha) = \frac{SR}{y_0}, \text{ and so } SR = y_0 \cos(2\alpha)$$

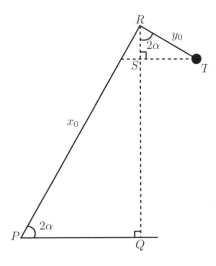

Figure 7.5

Then:

$$x_1 = PQ + ST$$
$$y_1 = QR - SR$$

So:

$$x_1 = x_0 \cos(2\alpha) + y_0 \sin(2\alpha)$$
$$y_1 = x_0 \sin(2\alpha) - y_0 \cos(2\alpha)$$

So our matrix for a reflection about the straight line through the origin which makes an angle of α with the positive x-axis is:

$$\begin{pmatrix} \cos(2\alpha) & \sin(2\alpha) \\ \sin(2\alpha) & -\cos(2\alpha) \end{pmatrix}$$

Let's check that this agrees with the results we've already seen for the angles 0 (i.e., reflection about the x-axis), $\frac{\pi}{2}$ (i.e., reflection about the y-axis) and $\frac{\pi}{4}$ (i.e., reflection about the line $y = x$):

$$\alpha = 0 : \begin{pmatrix} \cos(0) & \sin(0) \\ \sin(0) & -\cos(0) \end{pmatrix} \qquad = \begin{pmatrix} 1 & 0 \\ 0 & -1 \end{pmatrix}$$

$$\alpha = \frac{\pi}{2} : \begin{pmatrix} \cos(\pi) & \sin(\pi) \\ \sin(\pi) & -\cos(\pi) \end{pmatrix} \qquad = \begin{pmatrix} -1 & 0 \\ 0 & 1 \end{pmatrix}$$

$$\alpha = \frac{\pi}{4} : \begin{pmatrix} \cos\left(\frac{\pi}{2}\right) & \sin\left(\frac{\pi}{2}\right) \\ \sin\left(\frac{\pi}{2}\right) & -\cos\left(\frac{\pi}{2}\right) \end{pmatrix} \qquad = \begin{pmatrix} 0 & 1 \\ 1 & 0 \end{pmatrix}$$

That's exactly what we found at the beginning of the section! Hooray!

Let's Go Round Again

Another interesting linear map is rotation. Just like with reflection, we can find a matrix to express a general rotation through an angle θ anticlockwise about the origin. We'll be working with Figure 7.6 this time.

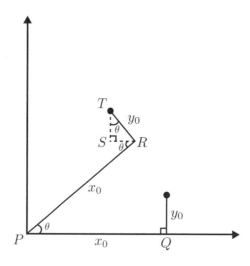

Figure 7.6

Once again we'll be using trigonometry to find our expressions for x_0 and y_0. First, we're going to do some geometry: PQ is parallel to SR, so $Q\hat{P}R = P\hat{R}S = \theta$. Then, because $P\hat{R}T$ is a right angle, we know that $S\hat{R}T = \frac{\pi}{2} - \theta$. Finally, $R\hat{T}S = \theta$ because RST is a triangle.

Now let's examine triangle PQR:

$$\sin\theta = \frac{QR}{x_0}, \text{ and so } QR = x_0 \sin\theta$$

$$\cos\theta = \frac{PQ}{x_0}, \text{ and so } PQ = x_0 \cos\theta$$

For triangle RST:

$$\sin\theta = \frac{SR}{y_0}, \text{ and so } SR = y_0 \sin\theta$$

$$\cos\theta = \frac{ST}{y_0}, \text{ and so } ST = y_0 \cos\theta$$

We see that:

$$x_1 = PQ - SR$$

$$y_1 = QR + ST$$

And so we arrive at:

$$x_1 = x_0 \cos \theta - y_0 \sin \theta$$
$$y_1 = x_0 \sin \theta + y_0 \cos \theta$$

So the corresponding matrix is:

$$\begin{pmatrix} \cos \theta & -\sin \theta \\ \sin \theta & \cos \theta \end{pmatrix}$$

Let's finish by taking a look at a specific example: rotation by π radians about the origin. The matrix for this would be:

$$\begin{pmatrix} \cos \pi & -\sin \pi \\ \sin \pi & \cos \pi \end{pmatrix} = \begin{pmatrix} -1 & 0 \\ 0 & -1 \end{pmatrix}$$

That's it for rotations: just don't forget that we rotate through θ radians *anti-clockwise* about the origin.

Eat Me, Drink Me

The last linear transformation that we're going to examine is enlargement. Up until now, it's likely that you'll have performed enlargements on various shapes in the plane, from a given centre of enlargement by a given scale factor. The kind of enlargement that we're going to do here will always have the origin as the centre of enlargement, but we *won't* be using the idea of a single "scale factor." Instead, we're going to allow a horizontal "stretch" by a given factor, and a vertical "stretch" by *another* scale factor. Of course, we can choose these two factors to be equal (resulting in the kind of enlargement you're probably already familiar with), but by allowing different parameters for the x and y stretch gives us greater freedom to transform points and shapes in the plane.

Just like with our other transformations, we're going to try to discover the matrix that allows us to perform the required enlargement. If we're stretching by a factor of m in the x direction and by a factor of n in the y direction (with both m and n strictly greater than 0), as in Figure 7.7, our system of equations will look like this:

$$x_1 = mx_0$$
$$y_1 = ny_0$$

So the corresponding matrix is:

$$\begin{pmatrix} m & 0 \\ 0 & n \end{pmatrix}$$

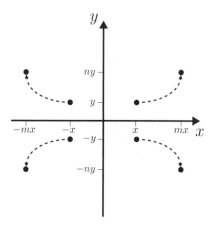

Figure 7.7

It's worth mentioning that the word "enlargement" can be a bit deceiving: if either of m or n are between 0 and 1 we actually get a "contraction" in that direction. If m or n are negative then we still get an enlargement, but also a change of sign. Think about the matrix multiplication if you're struggling to see why. Other than that, enlargement is fairly straightforward. After you remember what the matrix looks like, simply plugging in the required values of m and n will get you what you want!

Composition of Maps

In the previous section we explored the idea that matrix multiplication is exactly equivalent to the composition of the maps that the matrices represent: that is, performing one operation *and then* performing the other one on the result of the first. We're going to see a concrete example of this now that we have a good set of transformation matrices under our belts.

Let's imagine that we wanted to *first* reflect about the x-axis, and *then* to rotate by π radians about the origin. If we call the reflection A and the rotation B, we can find the corresponding matrices using the results we've just derived:

$$A = \begin{pmatrix} 1 & 0 \\ 0 & -1 \end{pmatrix}, \quad B = \begin{pmatrix} -1 & 0 \\ 0 & -1 \end{pmatrix}$$

We said that we want to perform the reflection first, and so using a matrix equation like we did in the last section gives us:

$$v_1 = Av_0$$

We then want to perform B on v_1, and so we get:

$$v_2 = Bv_1$$

Combining these two equations gives us:

$$v_2 = BAv_0$$

Look carefully at the order in which the matrices end up (remember, in general $AB \neq BA$ when we multiply matrices): the operation that we want to perform *first* goes on the right.

So what exactly is BA? Well, we simply multiply the matrices:

$$\begin{pmatrix} -1 & 0 \\ 0 & -1 \end{pmatrix} \begin{pmatrix} 1 & 0 \\ 0 & -1 \end{pmatrix} = \begin{pmatrix} -1 & 0 \\ 0 & 1 \end{pmatrix}$$

So there we have it: the matrix $\begin{pmatrix} -1 & 0 \\ 0 & 1 \end{pmatrix}$ is the *single* matrix corresponding to the transformation "first reflect in the x-axis, then rotate by π radians about the origin."

Being comfortable with this sort of calculation will be *vital* at university. Here's one last example before a set of exercises for you to try out for yourself:

> What is the matrix corresponding to the linear map "first enlarge by a factor of 3 horizontally and a factor of 2 vertically, and then rotate by an angle of $\frac{\pi}{2}$ anticlockwise about the origin"?

We need to find the two matrices for the individual transformations, and then multiply them together in the correct order. If we call the enlargement matrix C and the rotation matrix D, we have:

$$C = \begin{pmatrix} 3 & 0 \\ 0 & 2 \end{pmatrix} \text{ and } D = \begin{pmatrix} \cos\left(\frac{\pi}{2}\right) & -\sin\left(\frac{\pi}{2}\right) \\ \sin\left(\frac{\pi}{2}\right) & \cos\left(\frac{\pi}{2}\right) \end{pmatrix} = \begin{pmatrix} 0 & -1 \\ 1 & 0 \end{pmatrix}$$

Now we just need to put these matrices in the correct order and multiply them. We want to do C *then* D, which means that we need to perform the multiplication DC:

$$DC = \begin{pmatrix} 0 & -1 \\ 1 & 0 \end{pmatrix} \begin{pmatrix} 3 & 0 \\ 0 & 2 \end{pmatrix} = \begin{pmatrix} 0 & -2 \\ 3 & 0 \end{pmatrix}$$

There we have it: a single matrix to describe a composition of transformations.

EXERCISES

7.2.1. Find the matrix corresponding to the linear transformation "reflect all points in the straight line that makes an angle of $\frac{\pi}{6}$ with the positive x-axis".

7.2.2. Find the matrix corresponding to a reflection in the line $y = \sqrt{3}x$.

7.2.3. Find the matrix corresponding to the linear transformation "rotate all points $\frac{\pi}{4}$ radians anticlockwise about the origin."

7.2.4. Find the matrix corresponding to a rotation by $\frac{\pi}{3}$ radians anticlockwise about the origin.

7.2.5. What is the matrix corresponding to an enlargement by scale factor $\frac{1}{2}$ in both the x and y directions, with the origin as the centre of enlargement?

7.2.6. Find the matrix corresponding to an enlargement by a factor of 3 horizontally and a factor of 6 vertically.

7.2.7. Find the *single* matrix that corresponds to first rotating by $\frac{\pi}{6}$ radians anticlockwise about the origin, and *then* enlarging by a factor of 3 in both directions.

7.2.8. Find the *single* matrix that corresponds to first reflecting in the straight line that makes an angle of $\frac{3\pi}{4}$ radians with the positive x-axis, and *then* enlarging by a factor of 2 horizontally and a factor of 4 vertically.

7.2.9. Find the *single* matrix that corresponds to first rotating by $\frac{\pi}{2}$ radians anticlockwise about the origin, and *then* reflecting in the line $y = \frac{x}{\sqrt{3}}$.

7.2.10. Find the *single* matrix that corresponds to first reflecting in the x-axis, *then* rotating by $\frac{\pi}{3}$ radians anticlockwise about the origin, and *then* enlarging by a scale factor of 2 horizontally and a scale factor of 1 vertically.

7.3 Eigenvalues and Eigenvectors

It's fairly likely that eigenvalues and eigenvectors are new ideas to you. If you're already familiar with them it's safe to head over to the exercises right away, because we're going to spend this section exploring what eigenvalues and eigenvectors *are*, and also how to find them.

Genesis

So just what *are* eigenvalues and eigenvectors? Imagine that A is a $n \times n$ (i.e., square) matrix, and $\mathbf{v}$ is a vector with n components, but not all 0s. Finally, imagine that λ is a scalar (i.e., λ is "just a number"), so that the following equality holds:

$$A\mathbf{v} = \lambda\mathbf{v}$$

Then we say that λ is an *eigenvalue* of A, and that $\mathbf{v}$ is a *corresponding eigenvector*. It's definitely worth remembering that definition, even before exploring what each of the concepts means. It plays a huge part in lots of areas of mathematics (see "Where Now?" for an overview of just a couple!), and it's the kind of definition that is more than likely to come up in university exams.

Let's take a little time to look at why the equation is so special. It's saying that an eigenvector $\mathbf{v}$ of a matrix A is any vector which, upon performing the multiplication $A\mathbf{v}$, results in a vector that is some scalar multiple of $\mathbf{v}$ itself. A geometric interpretation of this is that the vector we get from the multiplication acts in the *same direction* as $\mathbf{v}$ (up to sign: it could be acting in the negative direction when compared to $\mathbf{v}$). This shows us why it's important that we can't allow $\mathbf{v}$ to be a vector of all 0s: if we allowed that to be called an eigenvector, then we would have to allow infinitely many values of λ, because the $\mathbf{0}$ vector has no magnitude (or "size").

Let's take a look at an example:

Let A be the matrix $\begin{pmatrix} 1 & 2 \\ 4 & 3 \end{pmatrix}$. A has an eigenvector $\mathbf{v} = \begin{pmatrix} 1 \\ -1 \end{pmatrix}$. What is the corresponding eigenvalue?

We know that we're going to be dealing with that all important equality $A\mathbf{v} = \lambda\mathbf{v}$, and this type of question is really just a case of plugging everything in. Here we go:

$$\begin{pmatrix} 1 & 2 \\ 4 & 3 \end{pmatrix}\begin{pmatrix} 1 \\ -1 \end{pmatrix} = \lambda\begin{pmatrix} 1 \\ -1 \end{pmatrix}$$

$$\begin{pmatrix} -1 \\ 1 \end{pmatrix} = \lambda\begin{pmatrix} 1 \\ -1 \end{pmatrix}$$

Here we can see that $\lambda = -1$, and so the answer to the question is that the corresponding eigenvalue is -1.

From Water to Wine

What we've done above is all well and good, but it isn't particularly *exciting* yet: the question told us an eigenvector, and we worked with it. But can we not be a little more fancy? Can we not work out the eigenvector for *ourselves?* Of course we can! Make sure you're up to speed with what it means for a matrix to be *invertible* (it means that the determinant of the matrix is not equal to 0) and bring a good eye for matrix algebra: we're going in.

Firstly, take the standard equation for eigenvalues and eigenvectors, and rearrange it so that we have the zero matrix on the right-hand side:

$$A\mathbf{v} = \lambda\mathbf{v}$$
$$A\mathbf{v} - \lambda\mathbf{v} = 0_n$$

Recall from the previous chapter that multiplying by the identity matrix won't affect anything. Let's do that now:

$$I_n A\mathbf{v} - \lambda I_n \mathbf{v} = I_n 0_n$$
$$A\mathbf{v} - \lambda I_n \mathbf{v} = 0_n$$

Factorising out a $\mathbf{v}$ (but being careful to keep things in the right order, because we know that order matters when we're dealing with matrix multiplication!) gives us:

$$(A - \lambda I_n)\,\mathbf{v} = 0_n$$

For now, let's *assume* that the matrix $(A - \lambda I_n)$ is invertible: that is, the determinant of $(A - \lambda I_n)$ is *not* equal to 0. If this is the case, then there exists a matrix $(A - \lambda I_n)^{-1}$ which is the inverse of the matrix $(A - \lambda I_n)$. Remember this point – we're going to be coming back here very soon!

Let's premultiply by this matrix:

$$(A - \lambda I_n)^{-1}(A - \lambda I_n)\,\mathbf{v} = (A - \lambda I_n)^{-1}\,0_n$$
$$\mathbf{v} = (A - \lambda I_n)^{-1}\,0_n$$

But *any* vector multiplied by the 0 matrix will result in a vector of 0s, and so we are left with the equality $\mathbf{v} = \mathbf{0}$. Hang on a minute: in our definition of an eigenvector, we *demanded* that $\mathbf{v}$ was not a vector of all 0s, and that's exactly what we've just ended up with! What went wrong? The *only* place that we did something questionable was when we stated that $(A - \lambda I_n)$ must be invertible, so we're going to need to adopt the view that $(A - \lambda I_n)$ is *not* invertible.

That means that we can write $\det(A - \lambda I_n) = 0$, and it is *this* equation that allows us to find eigenvalues for a matrix. We've got all the theory we need now, so let's get down to an example:

Find the eigenvalues of the matrix $\begin{pmatrix} 7 & -5 \\ 1 & 1 \end{pmatrix}$.

From what we've just seen above, we know that what we really need to be doing is solving the equation:

$$\det(A - \lambda I_2) = 0$$

Let's do it then!

If A is the matrix $\begin{pmatrix} a & b \\ c & d \end{pmatrix}$, then:

$$
\begin{aligned}
(A - \lambda I_2) &= \begin{pmatrix} a & b \\ c & d \end{pmatrix} - \lambda \begin{pmatrix} 1 & 0 \\ 0 & 1 \end{pmatrix} \\
&= \begin{pmatrix} a & b \\ c & d \end{pmatrix} - \begin{pmatrix} \lambda & 0 \\ 0 & \lambda \end{pmatrix} \\
&= \begin{pmatrix} a - \lambda & b \\ c & d - \lambda \end{pmatrix}
\end{aligned}
$$

Let's proceed with the specific example from the question. We need to solve the equation $\det(A - \lambda I_2) = 0$:

$$
\begin{aligned}
0 &= \det\left(\begin{pmatrix} 7 & -5 \\ 1 & 1 \end{pmatrix} - \lambda I_2 \right) \\
&= \det \begin{pmatrix} 7 - \lambda & -5 \\ 1 & 1 - \lambda \end{pmatrix} \\
&= (7 - \lambda)(1 - \lambda) + 5 \\
&= \lambda^2 - 8\lambda + 12 \\
&= (\lambda - 6)(\lambda - 2) \\
\lambda &= 2 \text{ and } 6
\end{aligned}
$$

There we have it: we've found the eigenvalues by *solving a quadratic equation!* For a 2×2 matrix, finding eigenvalues always boils down to this, so you can use a skill that you learned years ago to solve a whole new type of problem!

Any Vectors with That?

Now that we've seen how to find the eigenvalues for a matrix, all that remains is to find the eigenvectors for each of the eigenvalues. This is mostly just a case of solving simultaneous equations, but there are some things that we need to be careful of. For example, when solving $\det(A - \lambda I_n) = 0$, it's very possible that we end up with a quadratic equation that has repeated roots. We could end up with the equation $\lambda^2 + 6\lambda + 9 = 0$, which has the root $\lambda = -3$, but no other roots. What happens to the eigenvalues and the eigenvectors in this case?

We only get one eigenvalue rather than two as we would normally have, but this has a knock-on effect: When we calculate eigenvectors, we can end up with two *linearly independent* answers for our eigenvectors. In the context of a *pair* of eigenvectors, "linearly independent" means that one eigenvector is *not* a scalar multiple of the other: geometrically, they are vectors which are pointing in different (but not opposite!) directions. If you check our definitions of eigenvector and eigenvalue above, such a result *doesn't* contradict any of the theory that we developed there, but we're not going to delve any deeper into these types of matrices. Theory about the number of linearly independent eigenvectors for a given matrix (and what this can tell us about other things) may not appear in your degree until the end of the second year: instead, we're going to focus on the kind of example where we *can* find eigenvalues and eigenvectors without these sorts of problems.

As we said above, in "nice" examples all we need to do to find the eigenvectors from the eigenvalues is to solve simultaneous equations. Think back to the key equation:

$$A\mathbf{v} = \lambda\mathbf{v}$$

Given matrix A, after we've found the values of λ the only thing that we *don't* know in this equality is $\mathbf{v}$, and so we should certainly be able to calculate it. The best way to see how is to work through a couple of examples; that way we'll also be able to go over the process of finding eigenvalues again, because we always need to do that first. Let's revisit an example that we saw earlier, because that way we already know what part of the answer's going to be:

Find the eigenvalues and corresponding eigenvectors for the matrix $\begin{pmatrix} 1 & 2 \\ 4 & 3 \end{pmatrix}$.

We start off by solving the equation $\det(A - \lambda I_2) = 0$, like this:

$$\begin{aligned}
0 &= \det\left(\begin{pmatrix} 1 & 2 \\ 4 & 3 \end{pmatrix} - \begin{pmatrix} \lambda & 0 \\ 0 & \lambda \end{pmatrix} \right) \\
&= \det \begin{pmatrix} 1-\lambda & 2 \\ 4 & 3-\lambda \end{pmatrix} \\
&= (1-\lambda)(3-\lambda) - 8 \\
&= \lambda^2 - 4\lambda - 5 \\
&= (\lambda+1)(\lambda-5) \\
\lambda &= 5 \text{ and } -1
\end{aligned}$$

So we have the eigenvalues as -1 and 5. Now let's use the equation $A\mathbf{v} = \lambda\mathbf{v}$ to find the corresponding eigenvectors for each of the eigenvalues. We'll start with $\lambda = -1$. The eigenvector that we're trying to find will have two entries, and so we'll call it the vector $\begin{pmatrix} v_1 \\ v_2 \end{pmatrix}$:

$$A\mathbf{v} = \lambda\mathbf{v}$$

$$\begin{pmatrix} 1 & 2 \\ 4 & 3 \end{pmatrix} \begin{pmatrix} v_1 \\ v_2 \end{pmatrix} = -1 \begin{pmatrix} v_1 \\ v_2 \end{pmatrix}$$

$$\begin{pmatrix} v_1 + 2v_2 \\ 4v_1 + 3v_2 \end{pmatrix} = \begin{pmatrix} -v_1 \\ -v_2 \end{pmatrix}$$

Again, remember that we declare two matrices to be equal if *and only if* all of their components are equal. This means that we are dealing with the simultaneous equations:

$$v_1 + 2v_2 = -v_1$$
$$4v_1 + 3v_2 = -v_2$$

There are *infinitely many* solutions to this pair of equations: any vector that satisfies the condition $v_1 = -v_2$ will do. But that's *exactly what we should expect:* remember our geometric interpretation of an eigenvector involved the phrase "acts in the same direction as $\mathbf{v}$." By finding the relationship between v_1 and v_2 we find the *direction* in which our eigenvector points, but then it doesn't matter about the *magnitude* of the eigenvector, so long as it isn't equal to 0. It's generally best to pick a "convenient" solution to the eigenvector equation, so we'll choose $\begin{pmatrix} 1 \\ -1 \end{pmatrix}$, but we could have happily (and correctly!) chosen something like $\begin{pmatrix} 152361236 \\ -152361236 \end{pmatrix}$ or $\begin{pmatrix} -\pi \\ \pi \end{pmatrix}$.

Now that we've found the corresponding eigenvector for one of our eigenvalues, to find the corresponding eigenvector for the *other* eigenvalue we just repeat the process using *that* eigenvalue instead. Here goes:

$$A\mathbf{v} = \lambda\mathbf{v}$$

$$\begin{pmatrix} 1 & 2 \\ 4 & 3 \end{pmatrix} \begin{pmatrix} v_1 \\ v_2 \end{pmatrix} = 5 \begin{pmatrix} v_1 \\ v_2 \end{pmatrix}$$

$$\begin{pmatrix} v_1 + 2v_2 \\ 4v_1 + 3v_2 \end{pmatrix} = \begin{pmatrix} 5v_1 \\ 5v_2 \end{pmatrix}$$

Leaving us with the simultaneous equations:

$$v_1 + 2v_2 = 5v_1$$
$$4v_1 + 3v_2 = 5v_2$$

Both of these equations can be rearranged to give $2v_1 - v_2 = 0$, so any vector of the form $2v_1 = v_2$ is a solution. Specifically, we could choose the eigenvector

$\binom{1}{2}$. That's it: we're done! The answer is $\lambda = -1$, with corresponding eigen-vector $\binom{1}{-1}$, and $\lambda = 5$, with corresponding eigenvector $\binom{1}{2}$. We'll take a look at one more example before hitting the exercises.

Consider the matrix $A = \begin{pmatrix} 6 & 4 \\ -2 & -3 \end{pmatrix}$. What are its eigenvalues and corresponding eigenvectors?

$$\det (A - \lambda I_2) = (6 - \lambda)(-3 - \lambda) + 8 = 0$$
$$\lambda^2 - 3\lambda - 10 = 0$$
$$(\lambda + 2)(\lambda - 5) = 0$$
$$\lambda = 5 \text{ and } -2$$

Finding the corresponding eigenvector for $\lambda = 5$:

$$A\mathbf{v} = \lambda\mathbf{v}$$
$$\begin{pmatrix} 6 & 4 \\ -2 & -3 \end{pmatrix} \begin{pmatrix} v_1 \\ v_2 \end{pmatrix} = 5 \begin{pmatrix} v_1 \\ v_2 \end{pmatrix}$$
$$\begin{pmatrix} 6v_1 + 4v_2 \\ -2v_1 - 3v_2 \end{pmatrix} = \begin{pmatrix} 5v_1 \\ 5v_2 \end{pmatrix}$$

So our simultaneous equations are:

$$6v_1 + 4v_2 = 5v_1$$
$$-2v_1 - 3v_2 = 5v_2$$

Our solution vector is therefore any vector satisfying $v_1 = -4v_2$, so the vector $\binom{4}{-1}$ will certainly do. Now for the other eigenvalue:

$$A\mathbf{v} = \lambda\mathbf{v}$$
$$\begin{pmatrix} 6 & 4 \\ -2 & -3 \end{pmatrix} \begin{pmatrix} v_1 \\ v_2 \end{pmatrix} = -2 \begin{pmatrix} v_1 \\ v_2 \end{pmatrix}$$
$$\begin{pmatrix} 6v_1 + 4v_2 \\ -2v_1 - 3v_2 \end{pmatrix} = \begin{pmatrix} -2v_1 \\ -2v_2 \end{pmatrix}$$

So our simultaneous equations are:

$$6v_1 + 4v_2 = -2v_1$$
$$-2v_1 - 3v_2 = -2v_2$$

Our solution vector is therefore any vector satisfying $2v_1 = -v_2$, so the vector $\binom{1}{-2}$ will certainly do.

Putting everything together, our solution is $\lambda = 5$, with corresponding eigenvector $\binom{4}{-1}$, and $\lambda = -2$, with corresponding eigenvector $\binom{1}{-2}$.

EXERCISES

7.3.1. $\begin{pmatrix} 1 \\ -2 \end{pmatrix}$ is an eigenvector of $\begin{pmatrix} 3 & 1 \\ 2 & 2 \end{pmatrix}$. What is the corresponding eigenvalue?

7.3.2. $\begin{pmatrix} 1 \\ 1 \end{pmatrix}$ is an eigenvector of $\begin{pmatrix} 3 & 1 \\ 2 & 2 \end{pmatrix}$. What is the corresponding eigenvalue?

7.3.3. $\begin{pmatrix} 1 \\ -4 \end{pmatrix}$ is an eigenvector of $\begin{pmatrix} 6 & 2 \\ -8 & -4 \end{pmatrix}$. What is the corresponding eigenvalue?

7.3.4. $\begin{pmatrix} 1 \\ -1 \end{pmatrix}$ is an eigenvector of $\begin{pmatrix} 6 & 2 \\ -8 & -4 \end{pmatrix}$. What is the corresponding eigenvalue?

7.3.5. Consider the matrix $A = \begin{pmatrix} 4 & 3 \\ 2 & -1 \end{pmatrix}$. What are its eigenvalues and corresponding eigenvectors?

7.3.6. Consider the matrix $B = \begin{pmatrix} -2 & -1 \\ 7 & 6 \end{pmatrix}$. What are its eigenvalues and corresponding eigenvectors?

7.3.7. Consider the matrix $C = \begin{pmatrix} 1 & 5 \\ 1 & -3 \end{pmatrix}$. What are its eigenvalues and corresponding eigenvectors?

7.3.8. Consider the matrix $D = \begin{pmatrix} -5 & 3 \\ 3 & 3 \end{pmatrix}$. What are its eigenvalues and corresponding eigenvectors?

7.3.9. Consider the matrix $E = \begin{pmatrix} 7 & -1 \\ 4 & 2 \end{pmatrix}$. What are its eigenvalues and corresponding eigenvectors?

7.3.10. Consider the matrix $F = \begin{pmatrix} -9 & -10 \\ 3 & 2 \end{pmatrix}$. What are its eigenvalues and corresponding eigenvectors?

Where Now?

We hope that this chapter has been enlightening to you: seeing *why* we work with matrices in the way that we do is the starting point of something called "linear algebra". You'll probably meet linear algebra in your first year at university, and it leads to a huge array of courses in other areas of abstract algebra. If you want to do some background reading now, we'd recommend taking a look at *Elementary Linear Algebra, 9th Edition* (H. Anton and C. Rorres, John Wiley and Sons, 2005).

If you're already familiar with the transformation of translation (where we "shift" every point by a certain amount horizontally and vertically), then you might have wondered why we didn't include it in this chapter. The reason is that translation is *not* a linear map. If we go into the technical definition of a linear map, it actually demands that it doesn't matter which order we apply transformations in. For example, it doesn't make any difference if we enlarge something and then rotate or if we rotate it then enlarge it. We only made a fuss in the chapter about getting the matrix multiplication the right way around to really drive home the idea that AB doesn't necessarily equal BA in general matrix multiplication. We can see that translation doesn't satisfy this "order doesn't matter" property by imagining taking $(0,0)$ as our (x_0, y_0). If we first translate by $(1,0)$ and then enlarge by a factor of 2 in both directions, we end up at the point $(2,0)$. If, however, we do the enlarging first and *then* do the translation, we end up at the point $(1,0)$.

Finding the eigenvalues and eigenvectors for matrices comes up in a whole variety of situations. Finding equilibrium points in systems of differential equations is one: imagine there are two connected fish tanks with water flowing, through pipes, between them. One tank is initially filled with fresh water and the other is filled with very salty water. To find out what the concentration of salt in a given tank will be after a long while, we use eigenvectors! Another example of an application comes from statistics: if we're looking at a complicated random variable whose output is in two dimensions, we can use eigenvectors to make inferences about the random variable.

Of course, one very obvious way in which we can extend what we've done here is to consider matrices that are larger than 2×2. When we are dealing with bigger matrices, we need a more general approach to things like the determinant than we developed in the previous chapter. If we're looking at performing linear transformations in n dimensions, we'll definitely be needing square $n \times n$ matrices, and there's no reason why we should stop at $n = 2$. If you'd like to find out a little more, take a look at *Basic Linear Algebra* (T. Blyth and E. Robertson, Springer-Verlag, 1998).

8
Separable Differential Equations

Test Yourself

If you think you are already comfortable with this material, try these questions first and mark them using the answers at the back of the book. If you get them all right, you're probably ready to move straight on to the next chapter. If some look tricky, study the chapter first and then come back to these when you're ready.

1. Express the following as a single fraction:

$$\frac{3}{x+2} + \frac{x+5}{x^2}$$

2. Express the following as partial fractions:

$$\frac{7}{x^2 - 4}$$

3. Evaluate:

$$\int \frac{3}{x^2 + x} dx$$

4. Solve the following in y:

$$\frac{dy}{dx} = \frac{y^2}{x^2}$$

E. Hurst and M. Gould, *Bridging the Gap to University Mathematics*,
DOI: 10.1007/978-1-84800-290-6_8,
© Springer-Verlag London Limited 2009

5. Solve the following in y:

$$x\frac{dy}{dx} = yx^2$$

6. Solve the following in y:

$$4y^2\frac{dy}{dx} = \frac{y}{x}$$

7. Solve the following in y:

$$3x^2\frac{dy}{dx} = \frac{y}{x^2}$$

8. Solve the following expression. You need not rearrange your answer.

$$(x^2 + 3x)y\frac{dy}{dx} = 11$$

9. Solve the following expression. You need not rearrange your answer.

$$6\frac{dy}{dx} = 2xy^2 - 2xy$$

10. Solve the following expression. You need not rearrange your answer.

$$3(x^2 - 1)\frac{dy}{dx} = 6(y^2 + 2y)$$

8.1 Repetition Is the Key to Success

Reading this chapter, you'll fit into one of two categories. You will either be one of the people who have already studied a syllabus that deals with variable separable differential equations and partial fractions very thoroughly, or you will be one of the people that is completely new to at least one of these two topics because your syllabus didn't cover it. We remember that we started university as comfortable members of the first group, only to become aware that there were a good number of our peers who fell into the second. These people struggled with some of the concepts because the topics were dealt with very quickly, and so we thought it was wise to include them here. If you're comfortable with the ideas and the methods employed, you might want to stick around for some revision. Otherwise, head off to the next chapter. If you're new to the topics, then hopefully this chapter will give you a "stepping stone" up to learning the techniques more rigorously at university.

Partial Fractions

Imagine being faced with the following problem:

$$\text{Find} \int \frac{1}{x^2 - 1} \, dx.$$

At a first glance it seems innocent enough, but it is actually surprisingly difficult if we don't employ a certain trick. You see, the numerator is not a direct derivative of the denominator, and so we can't directly go to natural logarithms to solve it, nor can we hope for a quick success using a substitution. What we need to do is split the single fraction up into *partial fractions* and proceed from there.

"Partial fractions" is the name given to splitting a single fraction up into the sum of two or more fractions. Hopefully you're comfortable with the idea of combining two fractions into a single one, like this:

$$\frac{1}{x} - \frac{1}{x + 3} \equiv \frac{3}{x(x + 3)}$$

But what if we were given the fraction on the right-hand side, and asked to split it up into the sum of two fractions? How would we go about doing this? Let's work through this example to try to get the left-hand side of the expression back.

The first thing to notice is that the denominator of the combined fraction is simply the product of the denominators of the single fractions. That means that we can easily find the denominators of the two single fractions, but getting the numerators is a bit harder. Let's call them A and B. What we know is:

$$\frac{A}{x} + \frac{B}{x + 3} \equiv \frac{3}{x(x + 3)}$$

Now, if we try to combine our two single fractions in A and B, we get this:

$$\frac{A(x + 3) + Bx}{x(x + 3)} \equiv \frac{3}{x(x + 3)}$$

We know for certain that the denominators will always be the same (because we worked out what the single fraction denominators were by looking at the denominator of the combined fraction). This means that we only have to look at the numerators, and so we get the expression:

$$A(x + 3) + Bx \equiv 3$$

At this stage, there are two different methods that we can use to proceed. Firstly, we can use simultaneous equations to solve the problem – and indeed this is the method that we're going to stick with during the worked examples that we give. If we equate the constant coefficients, we get $3A = 3$, and so $A = 1$. Looking at the coefficients in x we have $A+B = 0$, and so by substituting $A = 1$ we arrive at $B = -1$. Setting these two values into the equation above gives us back:

$$\frac{3}{x(x+3)} \equiv \frac{1}{x} - \frac{1}{x+3}$$

We've succeeded in splitting the single fraction up into partial fractions.

But how about the *other* method that we mentioned? Well, that relies on the fact that the expression $A(x+3) + Bx = 3$ is actually an *identity*, and therefore must hold true for *any* value of x that we plug into it. This means that if we choose some nifty values for our x, we can actually simplify the problem. If we take $x = 0$ we can see that $3A = 3$, and so $A = 1$. Similarly, if we choose $x = -3$, we get $-3B = 3$, and so $B = -1$. Notice that these are exactly the same answers as we got from the simultaneous equations above. Decide which method you prefer and work at becoming familiar with it – as we said before, we're going to stick with the simultaneous equation method for our worked solutions.

Let's go back to the example with the integration, where we're trying to evaluate:

$$\int \frac{1}{x^2 - 1} dx$$

We know that the first goal is to split the denominator into two parts, and luckily $(x^2 - 1)$ can be expressed as the difference of two squares: that is $(x^2 - 1) = (x+1)(x-1)$, so we have the denominators of our two fractions as $(x+1)$ and $(x-1)$. Now, let's call the numerators A and B again, giving us the expression:

$$\frac{1}{x^2 - 1} \equiv \frac{A}{x+1} + \frac{B}{x-1}$$

Combining the two single fractions, we get the expression:

$$\frac{1}{x^2 - 1} \equiv \frac{A(x-1) + B(x+1)}{x^2 - 1}$$

From which we can deduce that $A(x-1) + B(x+1) \equiv 1$. We formulate our simultaneous equations by looking at the different coefficients in the expression – here, we have a coefficient in x (which is 0) and a constant coefficient (which is 1). So the simultaneous equations that we have are:

$$A + B = 0$$
$$B - A = 1$$

Make sure you're happy with where these equations come from, and then use any method that you're familiar with to solve them, to get the solution $A = -\frac{1}{2}$ and $B = \frac{1}{2}$. We can now plug these back into our original expression, to get:

$$\frac{1}{x^2 - 1} \equiv -\frac{1}{2(x-1)} + \frac{1}{2(x+1)}$$

Now we can solve the integral originally set out in the problem:

$$\int \frac{1}{x^2-1} dx = \frac{1}{2} \int \left(\frac{1}{x+1} - \frac{1}{x-1} \right) dx$$

$$= \frac{1}{2} \left(\ln|x+1| - \ln|x-1| \right) + k$$

$$= \frac{1}{2} \ln \left| \frac{x+1}{x-1} \right| + k$$

And there we have it: the quick guide to partial fractions.

And So. . .

There are a few cases where you realise that you can't find values for A and B, but these cases are reasonably rare and it's best to look them up as you come to them, as different anomalies have to be dealt with differently. Here we've only worked with examples where the denominators are linear, and if they aren't then things can change. There is a quick reference guide to some of the anomalies that can crop up with nonlinear denominators in the appendix of this book, so at a later date it may be worth taking a trip there and making up or finding some problems to practice with. For all other cases, where the A and B method is sufficient, here's a step-by-step guide for you:

- Look at the denominator of the fraction that you're trying to split into two.

- Find two things (which might be numbers or a function in x) that would multiply together to give this denominator.

- These two things are the denominators of your two single fractions. The numerators of these single fractions are A and B.

- Combine the two single fractions into a combined fraction in A and B.

- Compare this fraction with your original fraction to find A and B (which often means working through simultaneous equations).

- Substitute your values of A and B into your single fractions, and you're done!

Here's one last worked example before some exercises. This skill is crucial in all sorts of areas, so the exercise is jumbo-sized as a treat.

$$\text{Find} \int \frac{3}{2t^2 - t} dt.$$

Before we even think about the integration, let's start with some partial fraction action. The denominator here is $(2t^2 - t)$, which we can express as $t(2t - 1)$, and so the denominators of the single fractions are t and $(2t - 1)$. We now introduce A and B as the numerators, so we have:

$$\frac{3}{2t^2 - t} \equiv \frac{A}{t} + \frac{B}{2t - 1}$$

We know that our denominators are fine, so looking just at the numerators gives:

$$A(2t - 1) + Bt \equiv 3$$

Looking at the coefficients in t and constants, we get $A = -3$ and $B = 6$. This means that:

$$\int \frac{3}{2t^2 - t} dt = \int \left(\frac{6}{2t - 1} - \frac{3}{t} \right) dt = \frac{6}{2} \ln |2t - 1| - 3 \ln |t| + k$$

$$= 3(\ln |2t - 1| - \ln |t|) + k$$

$$= 3 \ln \left| \frac{2t - 1}{t} \right| + k$$

EXERCISES

8.1.1. Write the following as a single fraction:

$$\frac{3}{x + 2} + \frac{2}{5}$$

8.1.2. Write the following as a single fraction:

$$\frac{x + 3}{5} + \frac{2x - 1}{6}$$

8.1.3. Write the following as a single fraction:

$$\frac{x + 3}{2x - 1} + \frac{x^2}{x - 3}$$

8.1.4. Write the following as a single fraction:

$$\frac{x}{5} - \frac{2x}{6}$$

8.1.5. Write the following as partial fractions:

$$\frac{4x - 10}{x^2 - 2x - 8}$$

8.1.6. Write the following as partial fractions:

$$\frac{12x + 16}{x^2 + 2x - 3}$$

8.1.7. Write the following as partial fractions:

$$\frac{x + 17}{x^2 + x - 2}$$

8.1.8. Write the following as partial fractions:

$$\frac{2x + 8}{x^2 - 4}$$

8.1.9. Evaluate:

$$\int \frac{2x - 2}{x(x - 2)}\,dx$$

8.1.10. Evaluate:

$$\int \frac{7x - 5}{x^2 - 2x - 3}\,dx$$

8.1.11. Evaluate:

$$\int \frac{6x + 11}{x^2 - 3x - 4}\,dx$$

8.1.12. Evaluate:

$$\int \frac{13x - 55}{x^2 - 9x + 20}\,dx$$

8.2 Separation of Variables

Now we arrive at the heart of the chapter: variable separable differential equations. Separating variables is a procedure that we can use as part of the enormous "toolbox" of differential equations, in order to solve a specific type of problem.

A very basic differential equation problem might ask that we find the solution to:

$$\frac{dy}{dx} = 2x$$

Hopefully you are comfortable enough with this sort of problem to solve this "by sight" – that is, state that $y = x^2$ (plus some constant of integration). Throughout your previous studies you will have learnt lots of different tools to use in certain situations, such as the trigonometric identities or integration by parts, for example. Separation of varibles is a similar tool that lets us deal with problems that are given to us in the form:

$$\frac{dy}{dx} = R(x)S(y)$$

Firstly, let's decipher exactly what that means. The key thing to note is that $R(x)$ and $S(y)$ are all *functions* of the variables stated in the brackets. So to use separation of variables, we are looking for $\frac{dy}{dx}$ to be equal to some function of x times some function of y. Now, that might look like the sort of thing that never crops up, but remember that the number 1 is a function of x and a function of y, as is 0. Consider these expressions:

- $\frac{dy}{dx} = 4$
- $7x\frac{dy}{dx} = 1$
- $4\frac{dy}{dx} = 3y$
- $6xy\frac{dy}{dx} = 5$
- $2x\frac{dy}{dx} = (x^2 + 43)(y - 8)$

All of these are in the form that we require! As it turns out, separation of variables is actually a very widely used application.

So, what is the idea behind the separation of variables method? Well, put simply, it is the process of splitting the xs from the ys and then integrating both sides of the expression. Sound OK? After it has been mastered, it really isn't that different from the simplest of integration problems, but it does take some practice to get used to.

Terribly Sorry!

We must apologise, for we will now proceed to tell you lies. You see, $\frac{dy}{dx}$ is a symbol in its own right, and in normal calculation it is not acceptable to split it into two parts by multiplying through by dx. For example, we can't simply state that if $\frac{dy}{dx} = x$, then $dy = x \cdot dx$. What we *can* do, however, is to state that $\int dy = \int x dx$. There is a rigorous proof of this that you'll no doubt meet at university, but knowing it is not helpful in the quest to mastering how to solve such problems, and so we won't state it here. The lie that we will tell you in what follows is that we are "multiplying through by dx, and integrating." We are not doing this – it looks like we are, and in order to see the process at work this is a very helpful thing to imagine. Rest assured, however, that even though we are not doing what we *say* we are doing, the proof that we spoke about says that the *result* that is obtained from "multiplying through by dx, and integrating" is still valid: the only catch is that it should be arrived at via a much longer path.

So, with that in mind, let's proceed to working through an example to see separation of variables in action:

$$\text{Find a solution to } 2y\frac{dy}{dx} = 4x^3.$$

The first thing to check is that we have the required $\frac{dy}{dx} = R(x)S(y)$ form. With a quick rearrangement we get $R(x) = 4x^3$ and $S(y) = \frac{1}{2y}$, so we're all ready for liftoff. When tackling separation of variables problems, the aim is to get *all of the ys on the same side as the $\frac{dy}{dx}$, and all of the xs on the other side*. We do this by simply rearranging the expression: in this case it means putting it back how it was in the original statement of the question. When we're at this stage we can simply "multiply through by dx, and integrate" (remember about the lie!), and the problem is pretty much done. So here we go: "multiplying through by dx, and integrating":

$$\int 2y dy = \int 4x^3 dx$$
$$y^2 = x^4 + c$$
$$y = \sqrt{x^4 + c}$$

Hopefully that wasn't too scary: all that happens is that after "multiplying through by dx, and integrating," we rearrange to get an explicit solution in terms of y. If you're not too sure about any of the steps, take a closer look: be careful to note that when we're adding a constant of integration, we only need to do it *on one side*. If you prefer, you can think of this as adding c_1 to the left-hand side and c_2 to the right-hand side, then rearranging the expression to get $c_2 - c_1$ on the right-hand side and substituting $c = c_2 - c_1$, because we're dealing with *constants*. If you're happy then move on to this next (slightly more involved) example:

$$\text{Find a solution to } \frac{dy}{dx} = x^2 y, \text{ given that } y \neq 0.$$

Again we can see that our equation is in the required $\frac{dy}{dx} = R(x)S(y)$ form, so we may begin. Remember our preliminary goal in these problems:

Get all of the ys on the same side as the $\frac{dy}{dx}$, and all of the xs on the other side.

Unlike the last example, this *isn't* already done for us here. Doing it isn't really difficult at all, though: we have an extra y multiplied on the right-hand side, so all we do is divide through by a y (this time it isn't a lie, we really *are* allowed to divide through by a y, so long as we know $y \neq 0$). This gives us the expression:

$$\frac{dy}{dx} \cdot \frac{1}{y} = x^2$$

And now we have what we need: all of the ys with the $\frac{dy}{dx}$ and all of the xs on the other side. All that remains for us to do is to "multiply through by dx, and integrate," giving us:

$$\frac{dy}{dx} \cdot \frac{1}{y} = x^2$$

$$\int \frac{1}{y} dy = \int x^2 dx$$

$$\ln|y| = \frac{x^3}{3} + k$$

$$y = e^{\frac{x^3}{3} + k}$$

$$= A e^{\frac{x^3}{3}}$$

Hopefully you've seen the trick with changing our constant of integration to A before: all we're doing is exploiting the fact that $e^{\frac{x^3}{3} + k} = e^k \cdot e^{\frac{x^3}{3}}$, and then setting $A = e^k$.

So there we have it: a brand new skill for you to try out. So long as the original problem is in the $\frac{dy}{dx} = R(x)S(y)$ form we can use the separation of variables to find a solution, and the only key thing to remember is to get all of the ys on the same side as the $\frac{dy}{dx}$ and all of the xs on the other side. From then on, "multiplying through by dx, and integrating" gives a pretty standard problem to solve. Now it's time to try out your new weapon in the never-ending battle with integration:

EXERCISES

8.2.1. Solve the following in y:

$$\frac{dy}{dx} = 2x + 3$$

8.2.2. Solve the following in y:

$$\frac{dy}{dx} = \frac{x}{y}$$

8.2.3. Solve the following in y:

$$\frac{dy}{dx} = \frac{y}{x}$$

8.2.4. Solve the following in y:

$$\frac{dy}{dx} = xy$$

8.2.5. Solve the following in y:

$$\frac{dy}{dx}x = \frac{1}{y}$$

8.2.6. Solve the following in y:

$$\frac{dy}{dx}y^2 = 2xy$$

8.3 Combining the Tools

So far in this chapter, we've looked at partial fractions and variable separable differential equations. This may seem rather strange, as they have been two separate, unconnected ideas. However, if we look at the standard form of an equation which we tackle with separation of variables [i.e., $\frac{dy}{dx} = R(x)S(y)$], it becomes apparent that the two tools are part of the same arsenal.

We start with functions in x and y on both sides of the equation, and by the time we integrate we have only ys on the same side as the $\frac{dy}{dx}$ and only xs on the other side. In the section above, we could then simply proceed to the integration, but *these were very carefully chosen examples!* In reality, the "normal" thing to expect is an expression that requires the use of partial fractions, *at least* once, in order to proceed. Here's an example to illustrate:

Find a solution to $\dfrac{dy}{dx} = xy^2 + xy$

The first stages are as before, so let's plough straight through them:

$$\frac{dy}{dx} = xy^2 + xy$$
$$= x(y^2 + y)$$
$$\int \frac{1}{y^2 + y}dy = \int xdx$$

At this point, hopefully the difficulty is clear: while we have no problem integrating x, we can't directly integrate $\frac{1}{y^2+y}$. I'm sure you'll have guessed how we proceed: partial fractions. Here we go:

$$\frac{A}{y} + \frac{B}{y+1} = \frac{1}{y^2 + y}$$
$$A(y+1) + By = 1$$
$$A = 1$$
$$B = -1$$

From here, it's just business as usual, and so:

$$\int \left(\frac{1}{y} - \frac{1}{y+1}\right) dy = \int xdx$$
$$\ln|y| - \ln|y+1| = \frac{x^2}{2} + k$$
$$\ln\left|\frac{y}{y+1}\right| = \frac{x^2}{2} + k$$

At this point we're done. Depending on what the result is needed for, it may sometimes be beneficial to get rid of the logarithms – for example, if the equation is being used by an engineer to model a real situations, they may want to get rid of logarithms in order to make interpolation easier. For our purposes, however, the expression above is perfectly acceptable.

Double Trouble

Hopefully the previous example will prove to you that problems like this really aren't too scary, so long as you have the two individual skills (i.e., partial fractions and separation of variables) mastered. The only real complication that can rain on our parade is needing to use partial fractions twice: one final example will illustrate this better than we can explain it in words, but it really isn't anything extra to be afraid of.

Find a solution to $\dfrac{dy}{dx}3(x^2 + x) = 2(y^2 - 1)$

Again, the first few steps are standard, so I'll just write them out:

$$\frac{dy}{dx}3(x^2 + x) = 2(y^2 - 1)$$

$$\frac{dy}{dx}\frac{3}{y^2 - 1} = \frac{2}{x^2 + x}$$

$$\int \frac{3}{y^2 - 1}\,dy = \int \frac{2}{x^2 + x}\,dx$$

Hopefully you've spotted what's going on here: we now have trouble integrating *both* sides of the equation. How do we tackle this? We just put each side into partial fractions, and then go on as before. First, let's deal with the ys:

$$\frac{A}{y + 1} + \frac{B}{y - 1} = \frac{3}{y^2 - 1}$$

$$A(y - 1) + B(y + 1) = 3$$

$$B - A = 3$$

$$A + B = 0$$

$$B = \frac{3}{2}$$

$$A = \frac{-3}{2}$$

And now let's take a look at the xs:

$$\frac{C}{x} + \frac{D}{x + 1} = \frac{2}{x^2 + x}$$

$$C(x + 1) + Dx = 2$$

$$C = 2$$

$$D = -2$$

Once the partial fractions are sorted, we resume our scheduled broadcast:

$$\frac{3}{2}\int \left(\frac{1}{y - 1} - \frac{1}{y + 1}\right) dy = 2\int \left(\frac{1}{x} - \frac{1}{x + 1}\right) dx$$

$$\frac{3}{2}\ln\left|\frac{y - 1}{y + 1}\right| = 2\ln\left|\frac{x}{x + 1}\right| + k$$

And now for a commercial break:

EXERCISES

8.3.1. Solve the following:
$$\frac{dy}{dx} + 1 = y^2$$

8.3.2. Solve the following:
$$\frac{dy}{dx} + y = y^2$$

8.3.3. Solve the following:
$$\frac{dy}{dx}(x^2 - 3x) = \frac{6}{y}$$

8.3.4. Solve the following:
$$\frac{dy}{dx}(x^2 - 1) = \frac{7}{2y}$$

8.3.5. Solve the following:
$$\frac{dy}{dx} 5(x^2 - 3x) = 6(y^2 + y)$$

8.3.6. Solve the following:
$$\frac{dy}{dx} 4(2x^2 + 2x) = 9(y^2 - 1)$$

Where Now?

Partial fractions come in handy in all sorts of situations in mathematics, and so forming them is a skill that we get to use again and again. As we said in the section itself, there are rare cases where the standard A and B approach won't produce a solution: if you're keen to find out about this, there's some information in *A First Course in Differential Equations* (J. Logan, Springer, 2006).

In the grand scheme of things, the separation of variables is one of many tools that we can use when faced with calculus problems. It is only useful in certain situations, but it is a very helpful tool in the right circumstances. At university, you'll see the proof that we really can do what *looks* like "multiplying through by dx and integrating."

Separable differential equations are a great way to "feel" the mathematics of the real world in action. Lots of real-life processes can be modelled by equations of this type: *An Introduction to Ordinary Differential Equations* (J. Robinson, Cambridge University Press, 2004) is a great place to explore this kind of mathematics.

9

Integrating Factors

Test Yourself

If you think you are already comfortable with this material, try these questions first and mark them using the answers at the back of the book. If you get them all right, you're probably ready to move straight on to the next chapter. If some look tricky, study the chapter first and then come back to these when you're ready.

1. What form must an equation be in for us to make use of an integrating factor?

2. What is the integrating factor for the equation $\frac{dy}{dx} + 4x^3 y = 7x$?

3. What is the integrating factor for the equation $\frac{dy}{dx} + 2\frac{y}{x} = 14x^2$?

4. Find the integrating factor for the following expression, and multiply through by it:

$$2\frac{dy}{dx} + 10x^4 y = 6x^2$$

5. Rewrite the following expression in the form $\frac{d}{dx}(A(x)y) = B(x)$:

$$2x\frac{dy}{dx} + 2y = 17x^2$$

E. Hurst and M. Gould, *Bridging the Gap to University Mathematics*,
DOI: 10.1007/978-1-84800-290-6_9,
© Springer-Verlag London Limited 2009

6. Rewrite the following expression in the form $\frac{d}{dx}(A(x)y) = B(x)$:

$$e^{2x^3}\frac{dy}{dx} + e^{2x^3}6x^2y = 12$$

7. Find the integrating factor of the following equation and hence write it in the form $\frac{d}{dx}(A(x)y) = B(x)$:

$$3\frac{dy}{dx} + 6xy = 12x^3$$

8. Find a solution to the following equation by making use of an integrating factor:

$$\frac{dy}{dx} + \frac{y}{x} = 8$$

9. Find a solution to the following equation by making use of an integrating factor:

$$\frac{dy}{dx} + 2xy = 10x$$

10. Find a solution to the following equation by making use of an integrating factor:

$$\frac{dy}{dx} + \frac{y}{x} = e^x$$

9.1 Troubling Forms

As part of the long battle to solve differential equations, we develop an ever expanding "toolbox" of tricks that allows us to deal with more and more types of problems. One such trick is that of *integrating factors*, and throughout this chapter we're going to look at how these cunning fellows allow us to tackle a whole new range of questions.

This chapter is a little different from many of the others in that we're not going to "develop" the skill through increasingly difficult examples. The process of using integrating factors is quite lengthy and so what we're going to do is *build up* to solving problems through the three sections, but we won't actually have all of the tools in place to go ahead and *solve* problems until the final section. We apologise if this feels a little like wading in the dark, but the process is fairly complicated to get to grips with so it's better to be very accurate at all of the different stages than it is to muddle through problems before you're really ready.

Form and Function

The type of problems that can be solved using an integrating factor are all of a specific form. That is, we need the problem to be in the form:

$$\frac{dy}{dx} + P(x)y = Q(x)$$

Remember what the notation means: we need a $\frac{dy}{dx}$ term added to y lots of some function in x, and then we need all of that to equal some other function in x. Just as we saw in the previous chapter, although this looks like it would only happen on rare occasions, we can actually find differential equations of this form all over the place in mathematics.

Once we have identified that a problem is in the correct form, we can get to work. The basic idea of the integrating factors technique is to transform something that we can't directly integrate into something that we can, by multiplying through by some other function. In this section we're going to look at the way we find this other function – or to give it its proper title, *the integrating factor*.

Exponents and Multiplications

Although we're not going to look at a formal proof of *why* it is so, we can quickly and easily state what the integrating factor is. The proof is definitely valid, but in solving problems it is really of no use to us and so it is omitted here. Get ready – here comes the biggie:

> For an expression $\frac{dy}{dx} + P(x)y = Q(x)$, the integrating factor is $e^{\int P(x)dx}$.

Look carefully: that is *e to the power of* the integral of $P(x)$, with respect to x. In other words, we find the integral of $P(x)$ with respect to x, and then take this as the power of e. This may seem a little odd, but it most certainly works.

One quick thing to note is that if you hate constants of integration, you're in for a treat. When we do the integration at this stage, we totally ignore any constant of integration, and just put our e to the power of what we actually *get* in the process of integrating (because any constants would only "drop out" later in our working). With that in mind, let's look at a worked example:

> If $\frac{dy}{dx} + 4xy = 12$, what is the integrating factor?

Firstly, let's do a quick check that this expression is definitely in the form that we require. We need $\frac{dy}{dx} + P(x)y = Q(x)$, and so if we consider $P(x)$ to be $4x$ and $Q(x)$ to be 12, then we're ready to go. Now, remember what we need to do to find the integrating factor: We need to find $e^{\int P(x)dx}$. We've just identified $P(x)$ to be $4x$, and so the integrating factor here is going to be $e^{\int 4x dx}$. That's not too difficult to solve: remembering to ignore the constant of integration, we find that the integrating factor is e^{2x^2}. That's it! We're not yet concerned with how to use it, and the question only asked us to find the integrating factor, which is exactly what we've done. The answer to the problem is e^{2x^2}.

Hopefully that wasn't too painful to follow. Although we're going to leave the idea of *why* an integrating factor is useful to us until the next section, one thing that we can definitely do already is multiply through by our integrating factor so that we're ready to proceed with the problem. Here's an example that requests exactly that:

Find the integrating factor for the equation $2\frac{dy}{dx} + 4\frac{y}{x} = 12$, and then multiply both sides of the equation by it.

Again, our first task is to identify $P(x)$, so that we know what we're going to need to integrate. Here it's hidden slightly better: let's look at the form that the problem has to be in, so that we can try to find it.

We can see that the equation is not quite in the form $\frac{dy}{dx} + P(x)y = Q(x)$, because there is a 2 at the beginning. Before we go anywhere, we need to divide through by that 2 to get rid of it. The expression that we get from doing this is $\frac{dy}{dx} + \frac{2}{x}y = 6$. Comparing this to the standard form of $\frac{dy}{dx} + P(x)y = Q(x)$ shows us that $P(x)$ must be $\frac{2}{x}$.

Now we move on to the integration: We know that the integrating factor here must be $e^{\int \frac{2}{x} dx}$, which is $e^{2\ln x}$. At this stage we need to make use of a couple of tricks. First and foremost, we know that $2\ln x$ is the same as $\ln x^2$, and so we can write our integrating factor as $e^{\ln x^2}$. Secondly, $e^{\ln k}$ is simply k, for all values of k. This means that $e^{\ln x^2}$ is simply equal to x^2. Now that we've simplified the result of the integration, we can say that x^2 is the integrating factor.

The other thing that the question asks of us is to multiply through by the integrating factor. We do this in preparation for the next step, which we'll explore in the next section, but for now it's just a case of "doing as we're told." The result of multiplying the $\frac{dy}{dx} + P(x)y = Q(x)$ form of the expression through by the integrating factor is therefore $x^2\frac{dy}{dx} + 2xy = 6x^2$. So long as you've followed what's going on here, it's time for some exercises. If not, take another look – enlightenment shouldn't be too far away.

EXERCISES

9.1.1. If $\frac{dy}{dx} + P(x)y = Q(x)$, what is the integrating factor?

9.1.2. What is the integrating factor for the equation $\frac{dy}{dx} + 3y = 8x + 6$?

9.1.3. What is the integrating factor for the equation $\frac{dy}{dx} + 4xy = 16x$?

9.1.4. What is the integrating factor for the equation $3x\frac{dy}{dx} + 9x^2y = 1$?

9.1.5. What is the integrating factor for the equation $\frac{dy}{dx} + \frac{y}{x^3} = 7$?

9.1.6. What is the integrating factor for the equation $4x\frac{dy}{dx} + 8y = 16x^2$?

9.1.7. By expressing it in the $\frac{dy}{dx} + P(x)y = Q(x)$ form, find, and multiply through by, the integrating factor for the following equation:

$$2\frac{dy}{dx} + 2\frac{y}{x} = 12$$

9.1.8. By expressing it in the $\frac{dy}{dx} + P(x)y = Q(x)$ form, find, and multiply through by, the integrating factor for the following equation:

$$\frac{dy}{dx} + 3x^2y = 4x$$

9.2 Productivity

In the previous section, we've been to great lengths to find this elusive "integrating factor" character, and then to multiply through by it. But *why?* What is it doing? How is that helping? The answer lies in the products.

The Product Rule

Hopefully you'll already be familiar with the product rule for differentiation, which says that:

$$\frac{d}{dx}(M(x)N(x)) = M'(x)N(x) + M(x)N'(x)$$

But how does the product rule have anything to do with what we've just been up to? Well, as it turns out, by multiplying through by the integrating factor we've set ourselves up for a magnificent trick.

If we look at any of the above examples where we've multiplied through by an integrating factor, the left-hand side of the equation has a special property:

it looks just like what we'd expect to see if we'd used the product rule to differentiate some function. Let's pick up on the second of the worked examples from the previous section – we've already done all of the legwork and found the integrating factor, and after multiplying through by it we have:

$$x^2\frac{dy}{dx} + 2xy = 6x^2$$

Take a look at the left-hand side, where we have $x^2\frac{dy}{dx} + 2xy$. What function is this the derivative of? Well, in the first term we have $x^2\frac{dy}{dx}$, and in the second term we have $2xy$. Clearly the y is the thing that has been differentiated in the first term, and the x^2 is the thing that has been differentiated in the second, and so $\frac{d}{dx}(x^2y)$ would, by the product rule, give us $x^2\frac{dy}{dx} + 2xy$. This means that our original, enormous expression can now be written as $\frac{d}{dx}(x^2y) = 6x^2$. Amazing! It may take a little while to see this, so don't be disheartened if you can't spot it right away, but keep on looking and eventually it will suddenly slap you in the face. In a nice way.

Now that we've discovered this fantastic trick, you may be wondering what the purpose of using it is. Well, that will come in the next section, but for now just enjoy the excitement of the fact that we can take the original problem, re-arrange it, find an integrating factor, multiply through by it and then express the left-hand side as the derivative of some function! That is, the aim of what we're doing is to take some expression in the form $\frac{dy}{dx} + P(x)y = Q(x)$, and put it into the form $\frac{d}{dx}(A(x)y) = B(x)$. Here's another worked example with which to do exactly that – follow all of the steps carefully and ensure that you know what we're doing, when we're doing it:

By finding an integrating factor, express $\frac{dy}{dx} + 2xy = 20x$ in the form $\frac{d}{dx}(A(x)y) = B(x)$.

First up, it's integrating factor time. Hopefully you're well enough rehearsed in this to spot that it's $e^{\int 2x\,dx}$, which is e^{x^2}. Multiplying through by that gives us the expression:

$$e^{x^2}\frac{dy}{dx} + e^{x^2}2xy = 20xe^{x^2}$$

From here, we take the left-hand side and see that in the $e^{x^2}\frac{dy}{dx}$ term, a y must have been differentiated, and in the $e^{x^2}2xy$ term, an e^{x^2} must have been differentiated. This means that the left-hand side of the expression can be written as $\frac{d}{dx}(e^{x^2}y)$, and so the solution to the whole problem is just $\frac{d}{dx}(e^{x^2}y) = 20xe^{x^2}$. Again, as we said before, if you're not seeing that crucial "product rule" step, take a good look at the problem. When you think you've got what it takes, have a battle through these:

EXERCISES

9.2.1. Rewrite the following equation in the form $\frac{d}{dx}(A(x)y) = B(x)$:

$$3x\frac{dy}{dx} + 3y = 12x$$

9.2.2. Rewrite the following equation in the form $\frac{d}{dx}(A(x)y) = B(x)$:

$$4x^2\frac{dy}{dx} + 8xy = e^{x^2}$$

9.2.3. Rewrite the following equation in the form $\frac{d}{dx}(A(x)y) = B(x)$:

$$e^{3x}\frac{dy}{dx} + 3e^{3x}y = 16$$

9.2.4. Rewrite the following equation in the form $\frac{d}{dx}(A(x)y) = B(x)$:

$$e^{x^2}\frac{dy}{dx} + 2xe^{x^2}y = 8x$$

9.2.5. Find the integrating factor for the following equation, and hence write it in the form $\frac{d}{dx}(A(x)B(y)) = C(x)$:

$$\frac{dy}{dx} + 6xy = 4x$$

9.2.6. Find the integrating factor for the following expression, and hence write it in the form $\frac{d}{dx}(A(x)B(y)) = C(x)$:

$$2\frac{dy}{dx} + 4\frac{y}{x} = 2x^2$$

9.3 The Finishing Line

So we've done all this hard work and we've transformed an equation from the form $\frac{dy}{dx} + P(x)y = Q(x)$ into the form $\frac{d}{dx}(A(x)y) = B(x)$. Great. *But why?*

Well, once we have our equation in the form $\frac{d}{dx}(A(x)y) = B(x)$, we're so near the end of the problem that you can already sense the mass celebrations. You see, once we've reached the stage where $\frac{d}{dx}(A(x)y) = B(x)$, we can just integrate both sides of the equation and we have a solution. Integrating the right-hand side shouldn't be a problem because that's just a function, and now that we've got the left-hand side in the special form, integrating that is *even easier*, because we simply lose the $\frac{d}{dx}$ and we're done! Let's pick up on the last worked example that we left, where we had the expression as:

$$\frac{d}{dx}(e^{x^2}y) = 20xe^{x^2}$$

Integrating both sides:

$$\int \left(\frac{d}{dx}(e^{x^2}y) \right) dx = \int 20xe^{x^2} dx$$

Let's sort out the left side first, because that's so quick that if you blink you'll miss it:

$$e^{x^2}y = \int 20xe^{x^2} dx$$

Now for the right-hand side. Although integrating $20xe^{x^2}$ with respect to x may seem difficult, if we keep in mind that $2xe^{x^2}$ is the derivative of e^{x^2}, then we can quickly see that we can integrate the right-hand side to get $10e^{x^2}$ (take a quick look over this if you're not sure; it comes up often in these sorts of problems – remember that integrating is just the opposite to differentiating in this case).

Therefore, the solution to our whole problem is simply $e^{x^2}y = 10e^{x^2} + k$. Sadly, unlike earlier, we *do* need to include that constant of integration now, because we're doing a standard sort of integration, not a neat trick. But just as in the last chapter, we only need to add k to one side: we can think of it as adding k_1 to the left-hand side, k_2 to the right-hand side, and then letting $k = k_2 - k_1$. We can then tidy our answer up a bit to give a solution purely in terms of y. This is $y = 10 + ke^{-x^2}$.

And there we have it. It's taken a while, but we've made it all the way through a problem. Here's a guide to all the steps of the journey:

- Make sure that the initial expression is in the form $\frac{dy}{dx} + P(x)y = Q(x)$.

- Find the integrating factor, which is $e^{\int P(x)dx}$.

- Multiply through by it.

- Looking for the effects of the product rule, rewrite the left-hand side so that the expression is of the form $\frac{d}{dx}(A(x)y) = B(x)$.

- Integrate both sides with respect to x, remembering to include a constant of integration.

- If possible, rewrite the solution as y in terms of x.

With that list as our guide, here's one final worked example and some exercises as dessert.

$$\text{Solve } \frac{dy}{dx} + \frac{y}{x} = 7.$$

We have the form that we need, so all systems are go. The integrating factor here is $e^{\int \frac{1}{x} dx}$, which is $e^{\ln x}$, which is simply x. Multiplying through by the integrating factor gives:

$$x\frac{dy}{dx} + y = 7x$$

Rewriting this in the form $\frac{d}{dx}(A(x)y) = B(x)$ gives:

$$\frac{d}{dx}(xy) = 7x$$

Integrating both sides with respect to x:

$$\int \left(\frac{d}{dx}(xy)\right) dx = \int 7x dx$$

Performing the integration gives $xy = \frac{7x^2}{2} + k$, which we can tidy up to give the final solution as $y = \frac{7x}{2} + \frac{k}{x}$.

The only final thing that we may want to do is to use some initial conditions to find the constant of integration. If, for example, we were given the set of initial conditions $x = 1$, $y = 2$, we can substitute these in to our solution to get $2 = \frac{7}{2} + k$, and hence deduce that $k = \frac{-3}{2}$.

EXERCISES

9.3.1. Solve $\frac{dy}{dx} + \frac{y}{x} = 2$ by making use of an integrating factor.

9.3.2. Solve $x^2\frac{dy}{dx} + xy = x$ by making use of an integrating factor.

9.3.3. Solve $\frac{dy}{dx} + 3y = 4$ by making use of an integrating factor.

9.3.4. Solve $\frac{dy}{dx} + 2xy = 6x$ by making use of an integrating factor.

9.3.5. Solve $\frac{dy}{dx} + 4x^3y = 4x^3$ by making use of an integrating factor.

9.3.6. Solve $\frac{dy}{dx} + 3x^2y = 9x^2$, with initial conditions $x = 0, y = 0$, by making use of an integrating factor.

Where Now?

The integrating factor is a very useful method of getting an exact, algebraic answer to a differential equation. While iterative techniques on computers can now quickly give a numerical answer to many differential equation problems with initial conditions, finding an algebraic solution is both more useful and, of course, more satisfying!

One of the most famous uses of integrating factors is in Newton's law of cooling. You'll no doubt find out about this in your degree studies, but *An Introduction to Ordinary Differential Equations* (J. Robinson, Cambridge University Press, 2004) covers the subject very well, and has a good set of exercises to try. Newton's law of cooling is a very real, very useful concept, and it has a huge range of applications in the real world. Aside from telling you how long you have to wait for your coffee to be cool enough to drink, it is used by forensic experts in order to determine the time of death when they find a body!

10
Mechanics

Test Yourself

If you think you are already comfortable with this material, try these questions
first and mark them using the answers at the back of the book. If you get them
all right, you're probably ready to move straight on to the next chapter. If some
look tricky, study the chapter first and then come back to these when you're
ready.

1. At time $t = 0$, a particle is travelling with a velocity of $4\mathrm{ms}^{-1}$. The particle
    undergoes an acceleration of $2\mathrm{ms}^{-2}$ for 10 seconds. Use integration to find
    the velocity and displacement of the particle after the 10 seconds have
    passed.

2. In a scientific experiment, a particle is made to accelerate at $\frac{2}{t^2}\mathrm{ms}^{-2}$. After
    2 seconds have passed, it is recorded as having a velocity of $4 \mathrm{ms}^{-1}$ and
    a displacement of 2 m. Find the velocity and displacement after another 2
    seconds have passed.

3. What are Newton's three laws of motion?

4. What force would be required to make a particle of mass 5 kg accelerate
    at $3 \mathrm{ms}^{-2}$?

5. Throughout its short flight, a large firework is modelled as having a mass
    of $(20 - t)$ kg and travels at a velocity of $(3t^2 + 2) \mathrm{ms}^{-1}$. Find the force on
    the firework when $t = 5$.

E. Hurst and M. Gould, *Bridging the Gap to University Mathematics*,
DOI: 10.1007/978-1-84800-290-6_10,
© Springer-Verlag London Limited 2009

6. A sandbag of mass 40 kg is dropped from a hot air balloon. The drag coefficient of the bag is 8. What is the greatest speed that the bag could possibly reach, even if it were allowed to fall for an unlimited amount of time? Use the value $g = 9.8$ in your answer.

7. A box is thrown from the top of a high building and it reaches its terminal velocity of 15 ms^{-1}. The drag coefficient acting on the box is 20. Find the mass of the box.

8. A mass of 4 kg is suspended from the ceiling by two strings. The string on the left makes an angle of $60°$ with the horizontal, and the string on the right makes an angle of $30°$ with the horizontal. Find the tension in each string.

9. A box of mass 1 kg is placed on a slope, which makes an angle of $45°$ with the horizontal. The box is on the point of sliding down the slope. Find the magnitude of the frictional force holding the box in place.

10. A box sits on a truck's loading bay. When the driver activates the loading mechanism, the bay tilts at a rate of 1 degree per second. The mass of the box is 100 kg and the coefficient of friction is $\frac{1}{\sqrt{3}}$. How long will the driver have to wait before the box starts to slip down the loading bay?

10.1 Where You Want to Be

Many students going into university have some experience in mechanics. Whether this be through their studies in a specific mechanics module, from their studies in physics or from another source entirely, a good number of undergraduates won't be meeting mechanics for the first time at university. However, if you've never seen anything of the sort before, *don't despair* – you're probably not as behind as you think. Much of the mechanics before university has to do with working with specified formulae. Mechanics at university is *very different*: in your degree studies, you'll be much more focussed on deriving where equations come from, and looking at *why* things function as they do. So please don't worry – if you've never heard of acceleration before, or if you've studied mechanics pretty solidly over the past couple of years, this chapter is going to take what you're already familiar with and shape it so that you're ready to tackle the kind of tasks that you'll be meeting in your degree.

Acceleration, Velocity and Displacement

First up, we're going to take a look at some of the different ways that we can describe an object. In mechanics we mainly talk about *particles*, but don't worry too much about this if you're unfamiliar with the idea. For our purposes, all this means is that we treat objects as though they are perfectly rigid, are not affected by air resistance, and that all of their mass is concentrated in a single point. Basically, all we're saying is that in our calculations we're going to simplify the "real-world" situation just enough to be able to ignore some of the annoying factors that we would have to deal with if we did the experiments in real life.

When we're talking about a particle, three of the important factors that we need to describe are the particle's acceleration, a, its velocity, v, and its displacement, x. Let's now take a closer look at each of these ideas in turn:

- **Displacement:** if you've not studied mechanics before, this word is probably new to you. Displacement is simply the distance that something is from a given point, but it is a vector quantity and so it can be negative. For example, if a particle is 2 m to the right of the origin, we would say that its displacement from the origin is 2 m and its distance from the origin is 2 m. If, however, a particle is 2 m to the left of the origin, we would say that its displacement from the origin is -2 m, but its distance from the origin is still 2 m. The standard unit of displacement is metres.

- **Velocity:** this is the rate of change of displacement of the particle. In mechanics, we don't often use the word "speed," because speed does not include any notion of direction, whereas velocity can be either positive or negative, and is therefore more helpful to us. The standard unit of velocity is metres per second, ms^{-1}.

- **Acceleration:** this is the rate of change of velocity of the particle. If at time $t = 0$ the particle is travelling at 10 ms^{-1} and at time $t = 1$ the particle is travelling at 20 ms^{-1}, in one second the particle's velocity has increased by 10 ms^{-1}, so the acceleration is simply 10 ms^{-2}. Notice that the units of acceleration are ms^{-2}: metres per second squared. This is because what we're actually looking at is the change in metres per second, per second. Be sure to remember that acceleration is a *vector quantity:* it has both a magnitude and a *direction*.

You'll notice that we're very interested in using vector quantities in mechanics – that is, we need to be careful in deciding whether things are happening (be it displacement, velocity or acceleration) in the *positive* or in the *negative* direction. The good news is that when we first approach a problem in mechanics, we are free to *choose* which direction is the "positive" direction, so long as we remain consistent throughout the question. With mathematicians being the cunning fiends that they are, it comes as no surprise that people choose their "positive" direction to be the one that makes all of the subsequent calculations the *easiest* to compute.

Now we've got these definitions stated, we can start to examine some of the ideas in a more mathematical way. Looking again at the definition of acceleration, we say that this is the rate of change of velocity. In the example there, we had a time interval of 1 second and so the calculation was purely numerical. But how about if the time interval were much smaller? How about if it were infinitesimally small? Hopefully you see where we're headed: calculus.

When we say that acceleration is "the rate of change of velocity," this means that mathematically we must be dealing with the expression:

$$a = \frac{dv}{dt}$$

If we consider a time interval where t_0 is the start time and t_1 is the end time, we can integrate both sides with respect to t to obtain the very useful expression for "change in velocity," Δv:

$$\Delta v = \int_{t_0}^{t_1} a \; dt$$

Now, if we consider what Δv actually means, we can proceed to find one last useful formula. The expression Δv is the change in velocity, and so we can express this as being the velocity at the end of the interval minus the velocity at the beginning: $v_1 - v_0$. Substituting this into our equation and rearranging, we get:

$$v_1 = v_0 + \int_{t_0}^{t_1} a \; dt$$

We can generalise this statement a little further. In any problem, the starting velocity is a fixed number, but at any given moment the *present* velocity is actually a *function of time*. Think about that to make sure you're happy: velocity certainly varies with time, but once we've defined when t_0 is, whatever the velocity was at t_0 is a *fixed* quantity because *what's done is done*. This means that, for any t_n in the interval t_0 to t_1:

$$v(t_n) = v_0 + \int_{t_0}^{t_n} a \; dt$$

Second Order

Now that we've taken a look at acceleration in terms of velocity, we can go one step further. We defined velocity as being "the rate of change of displacement," and so using the same logic as we did with acceleration we arrive at the expression:

$$v = \frac{dx}{dt}$$

Now we have the opportunity to do something clever. We know that $a = \frac{dv}{dt}$ and $v = \frac{dx}{dt}$. How about *combining* those two statements? If we differentiate displacement we get velocity and if we differentiate velocity we get acceleration. Have you guessed where we're headed yet? The expression that we derive is:

$$a = \frac{dv}{dt} = \frac{d^2x}{dt^2}$$

Pretty handy, I'm sure you'll agree.

Now that we've got all of these beauties in place, all that remains between us and some examples is to look at one final expression. We found an expression for v_1 by first finding one for Δv, so is it possible to find one for x_1? Let's see what happens:

$$\frac{d^2x}{dt^2} = a$$

$$v(t_n) = v_0 + \int_{t_0}^{t_n} a \, dt$$

But remember that $v = \frac{dx}{dt}$, so:

$$\frac{dx}{dt}(t_n) = v_0 + \int_{t_0}^{t_n} a \, dt$$

Now let's integrate everything with respect to t_n:

$$\int_{t_0}^{t_1} \frac{dx}{dt}(t_n) \, dt_n = \int_{t_0}^{t_1} v_0 \, dt_n + \int_{t_0}^{t_1} \left(\int_{t_0}^{t_n} a \, dt \right) dt_n$$

$$x(t_1) - x(t_0) = \int_{t_0}^{t_1} v_0 \, dt_n + \int_{t_0}^{t_1} \left(\int_{t_0}^{t_n} a \, dt \right) dt_n$$

But $x(t_0)$ is "x at time t_0," which we also call x_0, and $x(t_1)$ is "x at time t_1," which we also call x_1. This notation makes everything look a little tidier:

$$x_1 - x_0 = \int_{t_0}^{t_1} v_0 \, dt_n + \int_{t_0}^{t_1} \left(\int_{t_0}^{t_n} a \, dt \right) dt_n$$

$$x_1 = x_0 + v_0(t_1 - t_0) + \int_{t_0}^{t_1} \left(\int_{t_0}^{t_n} a \, dt \right) dt_n$$

And there we have it, our final expression of this section. If you've done some mechanics dealing with "the SUVAT equations" before, doing the integration reveals a pleasant surprise: this whole expression yields the familiar friend $s = ut + \frac{1}{2}at^2$ when we have constant acceleration. If you'd like to see how, this is explored in the "Where Now?" section at the end of the chapter. Now for some worked examples of this crazy new stuff in action:

A particle is measured to be moving with an acceleration of 2 ms^{-2}, and at time $t_0 = 0$ its displacement is 5 m and its velocity is 3 ms^{-1}. Find the velocity and displacement of the particle after 10 seconds.

If you've done mechanics before, fight the urge to go to the SUVAT equations. These are a tool to help beginners. If you've never learned these equations, it's almost a blessing in disguise, because you won't be tempted to use them. Now that we're in with the big boys, we're using calculus all the way. Let's do the velocity first:

$$v_1 = v_0 + \int_{t_0}^{t_1} a \ dt$$

$$= 3 + \int_0^{10} 2 \ dt$$

$$= 3 + [2t]_0^{10}$$

$$= 3 + (20 - 0)$$

$$= 23\text{ms}^{-1}$$

And now for the displacement:

$$x_1 = x_0 + v_0(t_1 - t_0) + \int_{t_0}^{t_1} \left(\int_{t_0}^{t_n} a\ dt \right) dt_n$$

$$= 5 + 3(10 - 0) + \int_0^{10} \left(\int_0^{t_n} 2\ dt \right) dt_n$$

$$= 5 + 30 + \int_0^{10} \left([2t]_0^{t_n} \right) dt_n$$

$$= 35 + \int_0^{10} 2t_n\ dt_n$$

$$= 35 + \left[t_n{}^2 \right]_0^{10}$$

$$= 35 + (100 - 0)$$

$$= 135\text{m}$$

Hopefully you followed along – all that happened was integration. Don't be afraid to try doing this example again by yourself, and checking that you get the same answer: This kind of problem is *absolutely fundamental* to mechanics, and so it's a key skill to get to grips with.

When you're totally happy, here's another worked example, and then a set of exercises for you to tackle on your own:

A moving metal particle is in a magnetic field and undergoes an acceleration so that at time t, the acceleration of the particle is $\frac{6}{t^2}$. After 2 seconds have passed, the location of the particle is marked, and this point is labelled x_0. At this time, the velocity of the particle is 2 ms^{-1}. What is the velocity and displacement of the particle after another second passes?

The key difference between this question and the previous one is the way that the data is presented. Before, we got everything that we needed to know fed to us by the question, but here we need to extract our information from what we're told about the experiment. Firstly, we need to find the values of t_0 and t_1. This isn't too difficult: the measurement that we're given is taken after 2 seconds, and our answer is going to be about the particle after a further second has passed. This means that $t_0 = 2$ and $t_1 = 3$. Now, at t_0 we know that the velocity of the particle is 2 ms^{-1} and so we know that $v_0 = 2$. Finally, we have our expression for acceleration. In this instance it's dependent on time, but that doesn't change anything in our working so we simply state $a = \frac{6}{t^2}$. Now that we have extracted all of the numerical data that we need from the problem, let's go about solving it:

$$v_1 = v_0 + \int_{t_0}^{t_1} a \, dt$$

$$= 2 + \int_2^3 \frac{6}{t^2} dt$$

$$= 2 + \left[\frac{-6}{t} \right]_2^3$$

$$= 2 + \left(\frac{-6}{3} + \frac{6}{2} \right)$$

$$= 2 - 2 + 3$$

$$= 3 \text{ms}^{-1}$$

$$x_1 = x_0 + v_0(t_1 - t_0) + \int_{t_0}^{t_1} \left(\int_{t_0}^{t_n} a \, dt \right) dt_n$$

$$= 0 + 2(3 - 2) + \int_2^3 \left(\int_2^{t_n} \frac{6}{t^2} dt \right) dt_n$$

$$= 2 + \int_2^3 \left(\left[\frac{-6}{t} \right]_2^{t_n} \right) dt_n$$

$$= 2 + \int_2^3 \left(3 - \frac{6}{t_n} \right) dt_n$$

$$= 2 + [3t_n - 6\ln|t_n|]_2^3$$

$$= 2 + (9 - 6\ln 3) - (6 - 6\ln 2)$$

$$= (5 + 6(\ln 2 - \ln 3)) \text{m}$$

EXERCISES

10.1.1. At time $t = 0$, a particle moves with a velocity of 3 ms^{-1} and is accelerating at 5 ms^{-2}. Find the velocity of the particle at $t = 5$.

10.1.2. How far would the above particle travel in 10 seconds?

10.1.3. A particle has an initial velocity of 20 ms^{-1} and accelerates at a rate of 4 ms^{-2}. What velocity will the particle be travelling at after 3 seconds?

10.1.4. A bullet is travelling at 26 ms^{-1} when it enters some water and accelerates at a rate of -4 ms^{-2}. How far will the bullet travel through the water in 2 seconds?

10.1.5. A ball moving along a surface has an acceleration of $6t$. At time $t = 2$, the ball has a velocity of 2 ms^{-1}, and a displacement of 40 m. What velocity and displacement will it have at $t = 10$?

10.1.6. A charged particle moves through an energy field and undergoes an acceleration of $\frac{1}{t^2}$. The particle travels for 5 seconds and then is measured to have a velocity of 8 ms^{-1} and a displacement of 0. Find the velocity of the particle after a further 2 seconds have passed and the displacement of the particle 14 seconds after it started moving.

10.2 Faster! Faster!

I'm sure that at many times in your life, you'll have come across the name Isaac Newton. Newton really was the "inventor" of modern science, and a great deal of what we know today is thanks to the genius with the apple tree. In mechanics, one of Newton's most famous discoveries were his three laws of motion. Updated into the language of today, these are:

- **Newton's First Law:** A body will continue in uniform motion unless acted upon by a force.

- **Newton's Second Law:** Force equals rate of change of momentum.

- **Newton's Third Law:** For every action, there is an equal and opposite reaction.

While all of these laws are vitally important in their own right, we're going to look more closely at the second law. The key thing to notice here is that the law probably doesn't say what you thought it would. Most people *incorrectly* state the second law as $F = ma$. This *is* true in many cases, but it's not actually what the law is saying. If we blindly use $F = ma$, then we're actually making a simplification of the law, and in some cases this will give us an answer that is *just plain wrong*. Let's investigate this further.

Pitfalls and Temptations

Newton's second law, correctly stated, says:

$$F = \frac{dp}{dt}$$

where p is the letter that we use for momentum. If you've met the concept of momentum before you'll know that $p = mv$. If you haven't seen this before, don't worry – all we're saying is that the momentum of an object is equal to the object's mass multiplied by its velocity. Making the substitution $p = mv$ into Newton's second law yields:

$$F = \frac{d}{dt}(mv)$$

It's time to don those safety goggles. *This* is the step where Joe Public ploughs in and rearranges the equation to $F = m\frac{dv}{dt}$, and gets $F = ma$. But we're better mathematicians than your average bear, and so we're not going to do that *because it's wrong!* Moving the m term away from the instruction to differentiate is making the assumption that *mass is not a function of time*. If, in a certain problem where mass *is* constant, this is true – and this accounts for the "many times" when $F = ma$ *does* yield the correct solution. This is OK, but there are cases when the mass of the object will vary with time.

Consider an aeroplane. As it moves through the sky, it steadily uses up fuel. Not a phenomenal amount in a short space of time, but the mass of the aeroplane *does* decrease over the duration of a flight. The same is obviously true of cars and rockets as well. Imagine a snowball sliding down a large hill. As the snowball moves, it picks up more and more snow from the ground, and so its mass is increasing as it moves. At high velocities, when we need to take into account Einstein's work on relativity (don't worry, that's beyond the scope of this book!), the mass of *every* object is a function of time. Clearly, to make the assumption that the mass of a moving object is independent of time is *not good enough.*

How do we deal with this problem then? Well, in fact it's not too hard at all. Because *both* mass and velocity are functions of time, we have two functions in t that are multiplied together, and we need to differentiate. We need to call on a powerful weapon. You guessed it: the product rule.

If we differentiate using the product rule, we get the result:

$$F = m\frac{dv}{dt} + v\frac{dm}{dt}$$

Looking closely at what we have here is very encouraging. If we make the assumption that the mass of an object *is* independent of time, then the $\frac{dm}{dt}$ term takes the value 0. This is great, because then all we're left with is $F = m\frac{dv}{dt} + 0$: and *now* we're ready to make the move to $F = ma$ in these situations. If, however, we acknowledge the fact that the mass of an object is not independent of time, we still have the whole, correct expression to work with. Kind of makes you warm inside, doesn't it?

Terminal Velocity

Living our everyday lives on planet Earth, one of the things that we need to come to terms with quite quickly is the force of gravity. There's no escaping it, and at the surface it's always pulling down on us with around 9.8N kg^{-1}. This means for every kilogram of mass that you hold in your hand on the surface of the Earth, gravity will be pulling down with 9.8 Newtons of force. Let go of an object and, unsurprisingly, it falls. If we look at our equation $F = m\frac{dv}{dt} + v\frac{dm}{dt}$ we can see what happens. We know that every kilogram of mass is going to have a force of 9.8 Newtons acting on it, so the left hand side of the equation is simply going to be $9.8m$. If we assume that when we drop an object it doesn't lose any mass, we get the expression $9.8m = m\frac{dv}{dt}$.

If you've seen this before you won't be overwhelmed, but look at what happens: the ms cancel. We're left with $9.8 = \frac{dv}{dt}$: any object that is dropped, *regardless of mass*, will initially accelerate at 9.8 ms^{-2}.

In the previous section, where we were dealing with particles, we ignored the resistance to motion that drag forces pose. Now we're going to take a look at these drag forces, and see what happens. You see, from what we just saw, if there were no drag forces acting on a dropped falling object, it would simply keep accelerating. Assuming it was dropped from rest, after 1 second it would be travelling at a velocity of 9.8 ms^{-1}; after 2 seconds it would be travelling at a velocity of 19.6 ms^{-1}; after 100 seconds it would be travelling at a velocity of 980 ms^{-1}; and so on. Sadly, drag forces mean that things aren't as easy as that.

For a body moving through the air at high velocities, a good approximation of the drag forces acting upon it is the expression:

$$F_d = Dv^2$$

Here, F_d is the drag force, D is the drag coefficient (a constant related to physical factors such as the shape of the object and the material from which it is made) and v is the velocity of the object in question. Also, we saw above that the force of gravity acting on a body is always equal to mg, where m is the mass of the object and $g = 9.8$ ms^{-2}. Now, going back to our expression of Newton's second law, we can add in these extra factors. Drag *always* acts in the opposite direction to an object's motion, so in our calculations we have to *subtract* the effects it has from the motion that would occur if drag didn't exist. This gives us the expression:

$$mg - F_d = m\frac{dv}{dt} + v\frac{dm}{dt}$$

Now, we're going to again make the assumption that when we simply drop an object, it doesn't lose mass. This means that $\frac{dm}{dt} = 0$. Also making the substitution $F_d = Dv^2$, we have the expression:

$$mg - Dv^2 = m\frac{dv}{dt}$$

Dividing through by an m, we get:

$$\frac{dv}{dt} = g - \frac{Dv^2}{m}$$

This equation is very interesting to us, because of the fascinating way that the left- and the right-hand sides interact. If we release a body from rest, then it will initially accelerate at around g ms^{-2}, but the faster it falls, the stronger the drag (because of the v^2 in the drag term), and so the slower it accelerates. To really get your head around this, you need to make sure that you're totally happy with the concepts of acceleration and velocity.

Does this mean that there will be a point when the object will stop accelerating? *Yes, it does.* When the object is moving very quickly, the drag force will be equal in magnitude and opposite in direction to the downward pull of gravity. Looking at Newton's first law tells us that when this happens the object will continue to fall at a steady speed. It will be moving quickly, but there will be no more acceleration.

Examining this idea mathematically, we find that $\frac{dv}{dt} = 0$ when $g - \frac{Dv^2}{m} = 0$, which means when $g = \frac{Dv^2}{m}$. After a little bit of rearranging, we get the final expression:

$$v = \sqrt{\frac{mg}{D}}$$

We call this velocity the *terminal velocity* of the body. Interestingly, if somehow an object gets to be moving *faster* than its terminal velocity (say, for example, a bullet is fired downwards, so that its initial velocity is higher than its terminal velocity), the drag forces are stronger than the downward pull of gravity, and so the acceleration of the object is *negative*. This means the object keeps slowing down until it reaches – you've guessed it – terminal velocity.

EXERCISES

10.2.1. Given that a particle has a constant mass of 6 kg and accelerates at 5 ms^{-1}, find the force acting on the particle.

10.2.2. A 100 N force is applied to a constant mass of 4 kg. Find the acceleration of the mass.

10.2.3. A rocket has a mass of $(100 - 2t)$ kg, and travels at a constant velocity of 100 ms^{-1}. Find the force acting on the rocket.

10.2.4. As a meteorite burns up, it travels at a velocity of $(200t)$ ms^{-1} and has a mass of $(4t^2)$ kg. Find the force at $t = 2$.

10.2.5. If a falling body has a drag coefficient of 24.5 and a mass of 20 kg, what is its terminal velocity? Use the value $g = 9.8$ in your answer.

10.2.6. A skydiver has a mass of 80 kg, and notes that her terminal velocity is 2 ms^{-1}. What is her drag coefficient?

10.2.7. A rock is measured to fall with a terminal velocity of 10 ms^{-1} and has a drag coefficient of 0.5. What is the rock's mass?

10.3 Resolving Forces

Now that we've had a look at some objects in motion, we're going to take a look at some objects that are stationary. If we look back to Newton's first law, we can see that for an object to be still for more than just an instant all of the forces on it must have a total sum of 0: otherwise it would be accelerating. In this section, we're going to deal with two types of "stationary object" problems: those involving ropes and those involving slopes.

Ropes

Consider an object hanging on two different ropes from the ceiling in such a way that the object hangs still on the ropes and both ropes are taut. Why is such a situation possible? Why is the object not accelerating like *mad* when gravity is pulling it downwards and the ropes are both pulling on it too? We said that if a force acts on something, it will accelerate – *so what's going on?*

The answer to this question lies in resolving forces. The examples that we're going to look at won't be quite as complicated as the real world because we're going to stick to 2D problems. Don't worry, though: The concepts that we're dealing with in two dimensions are perfectly valid in the real, 3-dimensional world.

Take a look at Figure 10.1. It is a diagram of a particle suspended from the ceiling by two ropes.

Here, the forces in the ropes (we often call them "tensions") are acting in different directions, and yet the particle is stationary. Why? To solve a problem

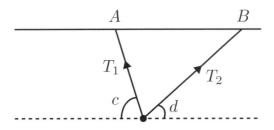

Figure 10.1

like this, we must look at the forces acting on the particle, and *resolve* them
into their components: horizontal and vertical. If we look *only* at the horizontal
components of the two forces first of all, we know that they must be equal and
opposite because the particle is not accelerating horizontally. If we then look
only at the vertical components of the forces in action (i.e., the gravitational
pull on the object, and the vertical components of the forces from the ropes), we
know that they must sum to 0 because the particle is not accelerating vertically.
We'll then get a pair of simultaneous equations that we can solve.

The first part of most "resolving forces" problems involves trigonometry. In
order to express the tensions T_1 and T_2 from the problem in their horizontal
and vertical components, we need to use sine and cosine. If we're going to look
at the horizontal direction first, then we get:

$$T_1 \cos c = T_2 \cos d$$

Now, if we look at the vertical direction, things are a tiny bit more complicated. We can't simply say that the two tensions are equal and opposite: they
aren't. Vertically, the sum of the two tensions is exactly equal and opposite to
the downward pull of gravity on the particle. Make sense? Don't worry about
the mass of the ropes themselves: in these sorts of problems, we make the assumption that the ropes don't actually have any significant mass of their own.
Anyway, resolving vertically:

$$mg = T_1 \sin c + T_2 \sin d$$

That's it. That's really all there is to it. If we're given values for some of these
variables we can go about getting numerical answers to these equations, but
otherwise what we've done is sufficient: we've formed two expressions (one horizontal and one vertical) for the situation in the diagram, and if someone were
to come along and tell us the necessary pieces of information, we could easily
tell them the tensions in the two ropes. Here's a worked example, doing exactly

that:

A 6 kg mass is to be suspended from the ceiling by two ropes. At the place where the particle hangs, rope 1 makes an angle of 60° from the horizontal, and rope 2 makes an angle of 30° from the horizontal, but on the opposite side of the object. Find the tensions in each of the ropes.

Reading the question, the set-up of this problem is identical to the diagram from before, but now we have specified angles and a specified mass. Making a labelled diagram is always an excellent way to start a problem like this, and Figure 10.2 is exactly the diagram that we need.

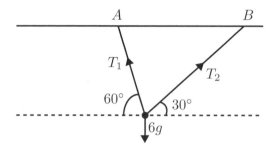

Figure 10.2

Starting with the horizontal first, we get:

$$T_1 \cos 60 = T_2 \cos 30$$
$$\frac{T_1}{2} = \frac{\sqrt{3}T_2}{2}$$
$$T_1 = \sqrt{3}T_2$$

And vertically:

$$6g = T_1 \sin 60 + T_2 \sin 30$$
$$= \frac{\sqrt{3}T_1}{2} + \frac{T_2}{2}$$
$$12g = \sqrt{3}T_1 + T_2$$

Finally, combining the two equations:

$$12g = 3T_2 + T_2$$
$$T_2 = 3g$$
$$T_1 = 3\sqrt{3}g$$

And there we have it. Read it again if you need to: all of the steps should be clear within a couple of runs through.

Slopes

Now that everything to do with ropes is clear, we can move on to problems dealing with stationary objects on slopes. We all know that if you place a block of wood on a steep enough slope, it will slide down it. But what about if we place that same block of wood on a gentle slope? Is there a chance that it won't move at all? *Yes, there is* – thanks to an old pal called friction. Let's try an example: have a look at Figure 10.3.

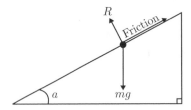

Figure 10.3

As the diagram shows, this system is a little more complicated than one dealing only with ropes. The particle is resting on an inclined surface, but because we make the assumption that the particle is very small, we also make the assumption that it cannot *roll* – otherwise, nothing would ever sit on an inclined plane!

There are three main forces that we need to look at in this type of problem:

- The **weight** force of the particle, mg.

- The **reaction** force of the surface on the particle, R.

- The **frictional** force.

As the diagram shows, the plane is inclined at an angle a from the horizontal. There are no other external forces in the problem.

Let's take a moment to think about the forces that are interacting here. The first (and most important) thing to notice is that the forces we have are not all parallel or perpendicular to each other, so we're going to need some trigonometry to make sense of the situation. Another key thing to note is that friction, just like drag in the previous section, always acts so as to directly oppose motion, and hence it acts directly up the plane. The reaction force of the plane on the particle (i.e., the force with which the plane pushes the particle) is *perpendicular to the plane*. This is a key fact: don't forget it!

The reaction force is perpendicular to the frictional force, and so we'll have to do the least amount of work if we resolve the weight force into its components parallel and perpendicular *to the plane*. This way we'll have all of the forces that we need expressed at right angles to each other, meaning that we can do some calculations to find exactly what we want.

So the key problem here is finding how we resolve the weight force, which acts directly downwards, into its components parallel and perpendicular to the plane. Luckily, as long as you can recreate Figure 10.4, that problem really isn't too difficult:

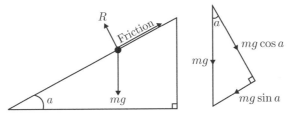

Figure 10.4

On the left is the original diagram, and on the right is the diagram of how we resolve that weight force. If you're confident with geometry, you might like to find a "similar triangles" argument as to why angle a ends up at the top in the new diagram, but if you're not too keen on that sort of thing then it's perfectly OK to memorise the diagram. This diagram is the key to solving all questions about particles on planes, so it's definitely worth the time investment of knowing this well enough to be able to draw it without really thinking. It comes up in every single question involving planes, so you're going to be using it over and over again.

Now that we have that sorted, we can start to form some equations. The forces R and $mg\cos a$ act along the same line, but in opposite directions. Because the particle does not sink into the plane or "jump up" from it, we know that these two forces must be equal. This gives us the equation:

$$R = mg\cos a$$

All that's left to do now is to look at the other pair of forces present in the system. Because our particle is not sliding up or down the plane, we know that the frictional force (we'll call it F_r) is equal to the component of the weight force that acts parallel to the plane. This yields the equation:

$$F_r = mg \sin a$$

And that's it. We have two equations into which we could substitute values in order to solve problems. Here's a worked example:

A particle of mass 10 kg is sitting on a plane inclined at 30° to the horizontal. It is not sliding down the plane. Find the magnitude of the frictional force acting on the particle.

Just like the problems involving ropes, our best chance at succeeding will come from a carefully drawn diagram, so take a look at Figure 10.5 now!

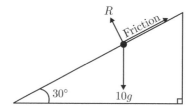

Figure 10.5

Now, we could formulate an equation in R, but that's not actually going to help with this problem, so we'll go straight for the equation for friction. We just saw that the equation to do with friction is simply $F_r = mg \sin a$, and so substituting the values into this equation will yield:

$$F_r = 10g \sin 30$$
$$= 5g$$

That really is all there is to it.

EXERCISES

10.3.1. A mass is sitting stationary on a smooth plane and is attached to two ropes. Each rope is horizontal, and the horizontal angle between the two ropes is 180°. The tension in one of the ropes is T. What is the tension in the other?

10.3.2. A 10 kg mass is suspended from the ceiling by two ropes. The rope on the left makes an angle of 45° with the horizontal and the rope on the right makes an angle of 45° with the horizontal. Find the tension in each of the ropes.

10.3.3. A 3 kg mass is suspended from the ceiling by two ropes. The rope on the left makes an angle of 60° with the horizontal and the rope on the right makes an angle of 30° with the horizontal. Find the tension in each of the ropes.

10.3.4. An 8 kg mass is suspended from the ceiling by two ropes. The rope on the left makes an angle of 45° with the horizontal and the rope on the right makes an angle of 30° with the horizontal. Find the tension in each of the ropes.

10.3.5. A particle of mass 10 kg is held by two ropes. The rope on the left is horizontal and the rope on the right makes an angle of 45° with the horizontal. If the tension in the horizontal rope is T, what is the tension in the other rope?

10.3.6. A particle of mass x kg sits on a slope which is inclined at $y°$ to the horizontal. Friction stops the particle from sliding. What is the magnitude of the frictional force?

Where Now?

The kind of problems that we've looked at in this chapter all require that we accept many simplifications of what really goes on in an experiment. The third section in particular has a lot of simplifications: the idea that there is no chance that the particle would topple or roll down the slope and the disregard for any drag forces other than friction are just two of many. As we develop more and more complicated models of what's really going on in the experiments, we need more and more complicated mathematics to be able to solve the problems.

What this chapter covers really is the tip of the iceberg. Classical mechanics like this could fill an entire book and still not deal with everything. Chapter 18 will give you the opportunity to look at a few more types of problem, but if you enjoy mechanics then your best bet is to go out and find some textbooks on it – you'll definitely have a good head-start on your degree if you're well prepared in this area. *University Physics, 11th Ed.* (H. Young and R. Freedman, Pearson, 2004) is a book we'd heartily recommend: it has detailed explanations and great examples to try out.

Think back to the derivation of the expression:

$$x_1 = x_0 + v_0(t_1 - t_0) + \int_{t_0}^{t_1} \left(\int_{t_0}^{t_n} a \, dt \right) dt_n$$

We promised that we'd show you how this expression "becomes" $s = ut + \frac{1}{2}at^2$ if we're working with constant acceleration (which will probably be familiar to you if you've already studied some mechanics). Here it is:

- Choose our initial frame of reference so that $t_0 = 0$ and $x_0 = 0$:

$$x_1 = v_0 \cdot t_1 + \int_0^{t_1} \left(\int_0^{t_n} a \, dt \right) dt_n$$

- We said that we're working with *constant* acceleration, so in this special case a is not a function of t. Integrating once yields:

$$x_1 = v_0 t_1 + \int_0^{t_1} a t_n \, dt_n$$

- Integrating again:

$$x_1 = v_0 t_1 + \frac{1}{2} a t_1^2$$

- In the "language" of the SUVAT equations, x_1 is the displacement, s; v_0 is the initial velocity, u; and t_1 is the finishing time, t:

$$s = ut + \frac{1}{2} a t^2$$

11
Logic, Sets and Functions

Test Yourself

If you think you are already comfortable with this material, try these questions first and mark them using the answers at the back of the book. If you get them all right, you're probably ready to move straight on to the next chapter. If some look tricky, study the chapter first and then come back to these when you're ready.

1. What does the symbol $\notin$ mean?

2. If $A = \{6x : x \in \mathbb{N}\}$ and $B = \{8x : x \in \mathbb{Z}\}$, is $-24 \in A \cup B$?

3. If $C = \{5x : x \in \mathbb{N}\}$ and $D = \{15x : x \in \mathbb{N}\}$, is $20 \in C \cap D$?

4. If $E = \{10x : x \in \mathbb{N}\}$ and $F = \{5x : x \in \mathbb{N}\}$, what are the three smallest strictly positive members of $F \setminus E$?

5. Write out the truth table for $\neg p$.

6. Write out the truth table for $p \vee q$.

7. Use a truth table to determine if $\neg(p \vee q)$ is logically equivalent to $(\neg p) \wedge (\neg q)$.

8. Use a truth table to determine if $(p \vee q) \wedge r$ is logically equivalent to $(p \wedge r) \wedge (q \wedge r)$.

E. Hurst and M. Gould, *Bridging the Gap to University Mathematics*,
DOI: 10.1007/978-1-84800-290-6_11,
© Springer-Verlag London Limited 2009

9. Is $f(x) = \sin x$, $f : \mathbb{R} \to \mathbb{R}$, injective, surjective, bijective or none of these?

10. Is $f(x) = x^3 + 7$, $f : \mathbb{R} \to \mathbb{R}$, injective, surjective, bijective or none of these?

11.1 Set Notation

Just like all areas of mathematics, working with sets requires that we have a solid, unambiguous language in which to communicate. Lots of academic books use set notation when they display logical arguments and proofs, and they are very helpful as a concise way to write answers to problems; we call these "solution sets." Despite their everyday use in the mathematical world, even the most basic usage of the notation surrounding sets is absent from most people's pre-degree education. For this reason, we're going to start at the beginning...

Dictionary Dialect

First and foremost, *what is a set?* Put simply, a set is an unordered collection of objects. It is important to remember that sets are unordered; an ordered version of a set is a vector. Not too bad, eh?

Unfortunately, the only way to learn most of the symbols and notation surrounding sets is to read them. Although we will discuss each one in a moment, here's the dreaded list to learn:

Symbol	Meaning
$x \in A$	x is an element of the set A
$y \notin B$	y is not an element of the set B
$M \subset N$	M is a proper subset of N
$P \supset Q$	Q is a proper subset of P
$M \subseteq N$	M is a subset of N
$P \supseteq Q$	Q is a subset of P
$A \cap B$	A and B
$A \cup B$	A or B (or both)
A^C	"Not A"
$A \setminus B$	The complement of A in B
$\emptyset$	The empty set
$\{\}$	"The set"
$\{x^2 : x \in \mathbb{Z}\}$	The set of squared xs, such that x is an integer
$\{x^2 \mid x \in \mathbb{Z}\}$	The set of squared xs, such that x is an integer

You might have already noticed a convention that we use: we give a set a capital letter and we give elements of a set a lowercase letter. Beyond that, each of the symbols obviously needs a little explanation before we can hope to use them accurately. Starting at the top, $\in$ and $\notin$ shouldn't be too much trouble, so let's go to an example straight away. If we say that A is the set of even numbers, then $2 \in A$ and $3 \notin A$. If we say that B is the set of roots of the polynomial $x^2 + 4x + 3$, then $-3 \in B$ and $7.63 \notin B$. Hopefully that's ample explanation.

Subsets are a bit more complicated. Even at degree level, the difference between the symbols $\subset$ and $\subseteq$ is often ignored, but there's no reason not to learn them properly first time around. The idea of a subset is simply this:

A subset is a set that contains some of, or all of, another set.

So the set of multiples of 4 is a subset of the set of even numbers. And the set of roots of the polynomial $x^2 - x - 6$ is a subset of all the integers greater than -10. Hopefully that's not too baffling.

A *proper subset* is simply a subset that *does not contain all of the original set*. Both the examples just given for "subset" are actually proper subsets. But consider this: If A is the set of all strictly positive even numbers and B is the set of multiples of 2 that are greater than 0, then A is a subset of B, but it is *not* a proper subset of B, because all of B is contained in A. A good trick to test whether you have a proper subset or not is this: If A is a subset of B *and* B is a subset of A (note this happens in our "even numbers" example), then A is not a proper subset of B, and B is not a proper subset of A. Otherwise, A is a proper subset of B.

To remember which way around the subset symbol goes, it might be helpful to think of $\subset$ as "contained in," so $A \subset B$ can be thought of "Set A is contained in Set B." As for which of $\subset$ and $\subseteq$ is for proper subsets, it's good to think of the analogy with the inequality symbols $<$ and $\leq$. While $<$ is "strictly less than," $\leq$ is "less than or equal to." Likewise, $\subset$ is a "strict" (i.e., proper) subset, and $\subseteq$ is "a subset, or equal to."

Although we use them in everyday life, the words "and," "or" and "not" have much more rigid definitions in mathematics. Let's tackle them one at a time:

- **And** — The word "and" is defined exactly how we use it in conversation. It demands that *both* things associated with it are true: for example, if you ask for fish and chips, you are saying "I would like fish" *and* "I would like chips." When talking about "and" in the context of sets, we use the symbol $\cap$, and we call it the "intersection'," so $A \cap B$ can be read as "A intersection B."

- **Or** — "Or" is the word that we have to be most careful with. The problem is, in everyday life we actually use the word quite sloppily. It is often hard to tell whether when someone says "or" they mean "either/or" or "and/or." For example, if I asked someone for "chips or potato wedges," I'd be perfectly happy if they brought me a plate of chips, a plate of potato wedges or a plate with half potato wedges and half chips. Yet if I said to someone that I would like "coffee or lemonade" I would be less than happy if they brought me a cup that contained a mixture of the two. The same "or" was used in both situations, yet one of the situations produces an unsatisfactory outcome — in the first situation, what I meant was "and/or," and in the second situation I meant "either/or." To avoid this difficulty in mathematics, it is necessary to have better clarity. For this reason, many years ago it was decreed that "or" in mathematics would be *inclusive or*: that is, "and/or." So if you are looking at the probability of A or B happening, you are looking at the probability of A happening, the probability of B happening and the probability of them *both* happening. If we don't want this "both" part on the end, then we have to specifically write "either/or" in our original statement: in this case, we would have written, "The probability of either A or B happening, but not both." The symbol for "or" is the symbol $\cup$ – often read as "union" in set theory.

- **Not** — Thankfully, "not" is an easy one. It simply negates the statement that it is attached to. In this book we denote "not" by writing a superscript C next to the event that we want to negate, so for the set "Not A" we would write A^C. There are some other ways to write "not," and different authors have different preferences: two common examples are to put a "bar" above the set (i.e., $\bar{A}$, pronounced "A bar") or to put an apostrophe after the set (i.e., A', pronounced "A prime"). (Too bad)^C, eh?

One last symbol before we try to take on whole expressions is $\setminus$. This is very similar to the "minus" sign – for example, $A \setminus B$ means "all of set A that is *not* in set B." An example should illustrate this: If A is the set of multiples of three and B is the set of multiples of five, then $A \setminus B$ is simply the set containing the numbers $3, 6, 9, 12, 18, 21, 24, 27, 33 \ldots$" and so on. If M is the set of even numbers and N is the set of multiples of four, then $M \setminus N$ is just the set $\{2, 6, 10, 14 \ldots\}$ Note that $A \setminus B$ is *not* the same as $B \setminus A$: in the two examples above, we see that $B \setminus A$ is $\{5, 10, 20, 25, 35, 40, 50 \ldots\}$ and $N \setminus M$ is actually "the empty set" – the set with no elements in it at all. We give this set the symbol $\emptyset$.

The Full Notation

Lastly, we come to full expressions in set notation – things like $\{x^2 : x \in \mathbb{Z}\}$. The key thing to remember is that anything in the curly brackets is in the set, the expression "on the left" is the property of the set, and the expression "on the right" is a restriction on the set. The thing that divides "the left" from "the right" is either a colon or a vertial line (: or |), depending on author preference: just like the table at the start of the chapter says, they mean exactly the same thing. When deducing what is meant by a set, reading : or | as "such that" makes good, logical sense of what is presented.

Therefore, something like $A = \{3x + 1 | x \in \mathbb{N}\}$ is simply read as, "The set of '$(3x+1)$'s, such that x is a natural number (recall from the chapter on complex numbers that the natural numbers are the numbers $1, 2, 3, \ldots$). Some elements of this set are $4, 7, 10, 13, 16 \ldots$..

One last example before the first set (no pun intended!) of exercises.

Find three elements of the set $G = \{5x : x \in \mathbb{Z}, x \neq 0\}$.

There are infinitely many members of this set, so simply choosing -30, 10 and 550 is perfectly good enough. One thing to note is that this time we have two restrictions on the set – the result is that $0 \notin G$, whereas it would be if we didn't have that second restriction.

EXERCISES

11.1.1. What does the symbol $\in$ mean?

11.1.2. What does the symbol $\setminus$ mean?

11.1.3. Describe in words what $\{4x | x \in \mathbb{Z}\}$ means.

11.1.4. If $A = \{3x : x \in \mathbb{N}\}$ and $B = \{5x : x \in \mathbb{Z}\}$, is $10 \in A \cup B$?

11.1.5. If $C = \{3x | x \in \mathbb{N}\}$ and $D = \{10x | x \in \mathbb{Z}\}$, is $20 \in C \cap D$?

11.1.6. If $E = \{6x : x \in \mathbb{N}\}$ and $F = \{3x : x \in \mathbb{N}\}$, how many elements of $E \setminus F$ are there that are less than 20?

11.1.7. If $G = \{2x + 1 | x \in \mathbb{N}\}$ and $H = \{3x | x \in \mathbb{N}\}$, what is the smallest *strictly positive* member of $G \setminus H$?

11.1.8. If $J = \{4x : x \in \mathbb{N}\}$ and $K = \{7x : x \in \mathbb{Z}\}$, what is the smallest *strictly positive* member of $J \cap K$?

11.2 Logical Equivalence

After our exploration of the symbols used in sets, it's time to start looking at some of the symbols used in *logical expressions*: You'll notice a remarkable similarity between these symbols and the symbols we've just looked at.

Consider a proposition p, which could either be true or false. We aren't saying anything specific about p, just that it's either true or it isn't. Then consider another proposition q, which is also (independently of p) either true or false. Recalling the definition of "and" given for sets in the last section, what combination of true and false would we require in order for "p and q" to be true? Hopefully you'll spot a link with the idea of "intersection": we'd need p to be true and q to be true. When working with logic, we write "and" as $\wedge$, which you'll see is strikingly similar to the symbol for intersection!

How about if we wanted "p or q" to be true? Again recalling the definition from the last section, this happens when p is true and q is false, when p is false and q is true, and when both p and q are true. The symbol for "or" looks (unsurprisingly!) similar to the symbol for "union": we write $\vee$ to mean "or."

As you've probabliy guessed, "not" has a logical symbol too. This one's a little different to the previous two, but it's still easy to remember. If we wanted to consider "not p" (which is the "negated" verion of p: that is, if p is true then "not p" is false and vice versa) we'd write $\neg p$.

The Truth

With the logical symbols in place, it's time to practise using them. At the start of this section we introduced the idea of arbitrary propositions p and q, and now we're going to look at putting individual propositions together to make new propositions. $\wedge$, $\vee$ and $\neg$ are examples of logical *connectives:* they can be used to combine component propositions to make a new proposition. This really isn't as confusing as it sounds: if p and q are both propositions then so too is $p \wedge q$: that is, individually p and q can either be true or false, and the proposition $p \wedge q$ can either be true or false. $p \wedge q$ is "true" when *both* p and q are true, and false otherwise. Remember that $\wedge$ means "and": go back and review the definition from the previous section if you need to.

Spotting whether a proposition like $p \wedge q$ is true or false isn't too difficult but as we deal with more difficult logical statements we need a more methodical approach. For this, we use a tool called *Truth Tables*.

A truth table is a table that looks at all of the possible "true or false" combinations between component propositions to see if a new proposition made out of these component propositions and logical connectives is true or false.

That definition is horribly theoretical, so here's an example:

Is the proposition $p \wedge q$ the same as the proposition $p \vee q$?

We know by definition that these are not the same, and so let's see that result using truth tables. Here is the truth table for $p \wedge q$:

p	$\wedge$	q
T	**T**	T
T	**F**	F
F	**F**	T
F	**F**	F
0	1	0

This probably looks like gibberish, so here's an explanation of what everything means. The logical statement that we're testing is along the top, divided up into its parts – nothing too astonishing there. The bottom line of the table is the order that we work through the table. In step 0, we write all of the possible "true or false" combinations of the propositions that we're considering. So for two propositions, we have:

- Both true

- p true and q false

- p false and q true

- Both false

These are the Ts and Fs in the columns of step 0. Then we move to the step 1 column, and fill in the gaps. The logical connective that we're looking at is $\wedge$, and so we ask the question, "Is $p \wedge p$ true?" Well, if both p and q are true then $p \wedge q$, or "p and q", is true. But if either one of p or q is false (or if they both are) then $p \wedge q$ is false. Now let's look at $p \vee q$:

p	$\vee$	q
T	**T**	T
T	**T**	F
F	**T**	T
F	**F**	F
0	1	0

Hopefully nothing there was a surprise to you: remember that $\vee$ means "or," and so $p \vee q$ will be true if either (or both) of p or q are true, but not if they are both false.

The question asked us to determine from these truth tables whether $p \wedge q$ and $p \vee q$ are the same thing. Well, two logical expressions are only the same if they are equivalent, and "equivalent" has a truth table too:

p	$\Leftrightarrow$	q
T	**T**	T
T	**F**	F
F	**F**	T
F	**T**	F
0	1	0

That is, two statements are equivalent if they "agree" where propositions are true and false. If you'd prefer to see that in action via an example, let's revisit the initial question by drawing one big truth table with stages 0, 1 and 2:

p	$\wedge$	q	$\Leftrightarrow$	p	$\vee$	q
T	**T**	T	**T**	T	**T**	T
T	**F**	F	**F**	T	**T**	F
F	**F**	T	**F**	F	**T**	T
F	**F**	F	**T**	F	**F**	F
0	1	0	2	0	1	0

You're probably wondering how we filled in this jumbo truth table. Well, we started by looking at the "stage 1" columns, and filling these with either "T" or "F" depending on whether the result of the test was true or false. These were precisely the results that we saw in the individual truth tables for $\vee$ and $\wedge$ above. After we've done that, we move on to the "stage 2" column, and fill it with either "T" or "F", drawing our inputs for this test *from the results in the "stage 1" columns.* From this table, because we can see that the final (in this case, "stage 2") column is *not* filled entirely with trues, then the two expressions $p \wedge q$ and $p \vee q$ are *not* the same.

Let's take a look at another example, and fill a truth table step-by-step:

Is $p \wedge (\neg(q \vee r))$ logically equivalent to the statement $(p \wedge (\neg q)) \wedge (p \wedge (\neg r))$?

The only difference here is that we're going to need more entries in step 0 to account for the fact that we have p, q and r to deal with. This means that there are eight possible "true" and "false" combinations to begin with.

Firstly, we need to work out the truth table for $\neg$, but that shouldn't be too hard:

$\neg$	p
F	T
T	F
0	1

See, nice and easy; if it's true, it becomes false. If it's false, it becomes true. OK, so here's the big truth table to answer the original question, shown step-by-step as it is filled in. Notice how the brackets tell us which operations are performed at each stage:

Step 0:

p	∧	(¬	(q	∨	r))	⇔	(p	∧	(¬	q))	∧	(p	∧	(¬	r))
T			T		T		T			T		T			T
T			T		F		T			T		T			F
T			F		T		T			F		T			T
T			F		F		T			F		T			F
F			T		T		F			T		F			T
F			T		F		F			T		F			F
F			F		T		F			F		F			T
F			F		F		F			F		F			F
0			0		0		0			0		0			0

Step 1:

p	∧	(¬	(q	∨	r))	⇔	(p	∧	(¬	q))	∧	(p	∧	(¬	r))
T			T	**T**	T		T		**F**	T		T		**F**	T
T			T	**T**	F		T		**F**	T		T		**T**	F
T			F	**T**	T		T		**T**	F		T		**F**	T
T			F	**F**	F		T		**T**	F		T		**T**	F
F			T	**T**	T		F		**F**	T		F		**F**	T
F			T	**T**	F		F		**F**	T		F		**T**	F
F			F	**T**	T		F		**T**	F		F		**F**	T
F			F	**F**	F		F		**T**	F		F		**T**	F
0			0	1	0		0		1	0		0		1	0

Step 2:

p	∧	(¬	(q	∨	r))	⇔	(p	∧	(¬	q))	∧	(p	∧	(¬	r))
T		**F**	T	T	T		T	**F**	F	T		T	**F**	F	T
T		**F**	T	T	F		T	**F**	F	T		T	**T**	T	F
T		**F**	F	T	T		T	**T**	T	F		T	**F**	F	T
T		**T**	F	F	F		T	**T**	T	F		T	**T**	T	F
F		**F**	T	T	T		F	**F**	F	T		F	**F**	F	T
F		**F**	T	T	F		F	**F**	F	T		F	**F**	T	F
F		**F**	F	T	T		F	**F**	T	F		F	**F**	F	T
F		**T**	F	F	F		F	**F**	T	F		F	**F**	T	F
0		2	0	1	0		0	2	1	0		0	2	1	0

Step 3:

p	∧	(¬	(q	∨	r))	⇔	(p	∧	(¬	q))	∧	(p	∧	(¬	r))
T	**F**	F	T	T	T		T	F	F	T	**F**	T	F	F	T
T	**F**	F	T	T	F		T	F	F	T	**F**	T	T	T	F
T	**F**	F	F	T	T		T	T	T	F	**F**	T	F	F	T
T	**T**	T	F	F	F		T	T	T	F	**T**	T	T	T	F
F	**F**	F	T	T	T		F	F	F	T	**F**	F	F	F	T
F	**F**	F	T	T	F		F	F	F	T	**F**	F	F	T	F
F	**F**	F	F	T	T		F	F	T	F	**F**	F	F	F	T
F	**F**	T	F	F	F		F	F	T	F	**F**	F	F	T	F
0	3	2	0	1	0		0	2	1	0	3	0	2	1	0

Step 4:

p	$\wedge$	$(\neg$	$(q$	$\vee$	$r))$	$\Leftrightarrow$	$(p$	$\wedge$	$(\neg$	$q))$	$\wedge$	$(p$	$\wedge$	$(\neg$	$r))$
T	F	F	T	T	T	T	T	F	F	T	F	T	F	F	T
T	F	F	T	T	F	T	T	F	F	T	F	T	T	T	F
T	F	F	F	T	T	T	T	T	T	F	F	T	F	F	T
T	T	T	F	F	F	T	T	T	T	F	T	T	T	T	F
F	F	F	T	T	T	T	F	F	F	T	F	F	F	F	T
F	F	F	T	T	F	T	F	F	F	T	F	F	F	T	F
F	F	F	F	T	T	T	F	F	T	F	F	F	F	F	T
F	F	T	F	F	F	T	F	F	T	F	F	F	F	T	F
0	3	2	0	1	0	4	0	2	1	0	3	0	2	1	0

There we have it: at the end of step 4, we see that we get a whole column of trues. Even though the statements do look rather different, they are in fact logically equivalent.

EXERCISES

11.2.1. Write out the truth table for $p \Leftrightarrow q$.

11.2.2. Write out the truth table for $p \wedge q$.

11.2.3. Use a truth table to determine if $p \wedge (\neg q)$ is logically equivalent to $\neg(p \wedge q)$.

11.2.4. Use a truth table to determine if $p \vee (q \wedge r)$ is logically equivalent to $(p \vee q) \wedge (p \vee r)$.

11.2.5. Use a truth table to determine if $\neg(p \vee q)$ is logically equivalent to $(\neg p) \vee (\neg q)$.

11.2.6. Use a truth table to determine if $\neg(p \wedge q)$ is logically equivalent to $(\neg p) \vee (\neg q)$.

11.3 Functions

There is also a fair amount of logical discovery to be made with functions. Sadly, most of it is quite complicated so what we're going to do here is to look at some definitions. That way, when the time comes to study functions properly you'll have this stuff firmly under your belt.

Needles and Knighthoods

Firstly, let's answer the basic question: what *is* a function? We often draw graphs *of* functions, but what is a function to begin with? Well, quite simply, a function is something that relates every input to *exactly one* output. This means if we put a number into a function, then we'll never get two or more outputs, and so our familiar friends like x^2, $\sin x$ and the like are all indeed functions. But the line shown in Figure 11.1 is not, because at the value marked with the dotted line there are two output values defined for the given input.

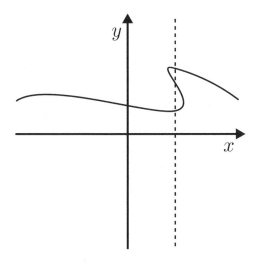

Figure 11.1

Keeping this visual approach in mind should help you determine whether things are functions or not quickly and easily: if you can find any value of x that has more than one value of $f(x)$, then you don't have a function.

When we define a function, we need to describe two things about it: what it *does* to things that we put through it, and also where it maps from and to. Most functions you'll have met so far map from the real numbers to the real numbers – that is, we can put any real number in and we will get a real number out. But not all functions work this way: imagine the function $f(x) = \sqrt{x}$, where we take the *positive* square root (so, for example, we say $f(9) = 3$ and *not* $f(9) = \pm 3$). We definitely can't have this function going from the reals to the reals, because what would we do when we try to apply this function to a negative number? Here, we have two options: we can either restrict the *domain* (i.e., the set of numbers that go *in* to the function), or we can expand the *range* (i.e., the set of numbers that come *out* of the function). This means either of the functions $f(x) = \sqrt{x}, f : \mathbb{R}_{\geq 0} \to \mathbb{R}$ or $f(x) = \sqrt{x}, f : \mathbb{R} \to \mathbb{C}$ are both safe to use (where $\mathbb{R}_{\geq 0}$ is the notation for "the non-negative real numbers" and $\mathbb{C}$ is the set of complex numbers).

It's *essential* that you keep the *order* of the letters in the definition of "function" in mind. It's convenient to use $y = f(x)$, as you probably already have done for many years.

If we have only one value of y for each value of x, we have a function. But if we know that we have a function and we *also* know that we have only one value of x for each value of y, then our function is *injective*. Look carefully at the way the letters go around in that definition: for a function to be injective, we *reverse* the roles of x and y in the definition of "function." So the function $y = f(x) = x$ is injective, because each value of y only has one value of x, but the function $y = f(x) = x^2$ is *not* injective, because at, say, $y = 4$, we have both $x = 2$ and $x = -2$ mapping there. A less formal name for injectivity is "one-to-one": This comes from the idea that for a function to be injective, a point in the range that is mapped to will be mapped to only once.

As well as injectivity, there is another property that functions can have: surjectivity. A function is *surjective* if it maps to all possible values of the range. This means we can choose any value of y in our given range and find at least one value of x (in our domain) that is mapped to it. For example, the function $f(x) = x, f : \mathbb{R} \to \mathbb{R}$ is surjective because every value of y will have a value of x mapped to it, but the function $f(x) = x^2, f : \mathbb{R} \to \mathbb{R}$ is *not* surjective because negative values of y will never be mapped to. A less formal name for surjectivity is "onto": this comes from the idea that wherever you choose in the range, you will be "mapped onto." We can also write the definition of surjectivity in formal language:

> If $f : A \to B$, and for every $b \in B$ there exists some $a \in A$ such that $f(a) = b$, then f is surjective.

The one final property of functions that we're going to look at is somewhat easier. Here it is:

> If a function is both injective and surjective, then it is *bijective*.

Thankfully, that's a lot easier to remember than the other two! Being equipped with a good knowledge of these three properties that functions may or may not have is excellent preparation for your degree. These properties come up *all over the place* in proofs, and so getting a firm grasp of them now is most definitely going to be beneficial. Learning them thoroughly is going to take a while, but be persistent! It might also help if you think about some functions that you already know (e.g. x^4 or $\cos x$), and work out which of the properties they have.

EXERCISES

11.3.1. Write, in words, the definition of "surjective."

11.3.2. Write, in words, the definition of "injective."

11.3.3. If $f : A \to B$, and for every $b \in B$ there exists $a \in A$ such that $f(a) = b$, *and* $a = b \Rightarrow f(a) = f(b)$, is f bijective?

11.3.4. Is the function $f(x) = 3x$, $f : \mathbb{N} \to \mathbb{Z}$ surjective?

11.3.5. Is the function $f(x) = \frac{x}{2}$, $f : \mathbb{Z} \to \mathbb{Q}$ injective?

11.3.6. Is the function $f(x) = 5x$, $f : \mathbb{N} \to \mathbb{Z}$ bijective?

11.3.7. What is the problem with the function $f(x) = \frac{x}{2}$, $f : \mathbb{Z} \to \mathbb{Z}$?

11.3.8. How many of $\sin x$, $\cos x$ and $\tan x$ are surjective, if $f : \mathbb{R} \to \mathbb{R}$?

Where Now?

Learning to use the language of sets and functions is obviously only a starting block from which we can access the "real" mathematics. This language is the everyday language of all mathematicians' work in logic, and so being fluent in understanding it is absolutely crucial.

If you'd like to know a little about using what we've learned here, *Algebra and Geometry* (A. Beardon, Cambridge University Press, 2005) has a good exploration of permutations on sets (which makes good use of concepts such as bijectivity).

In analysis, lots of "disproof by contradiction" arguments work around injectivity or surjectivity. We often start a proof by assuming a function to be either one or the other, doing some fancy tricks, and concluding that it *isn't* what we said it was – hence disproving the initial statement. In fact, this sort of proof is central to many areas of degree mathematics.

12

Proof Methods

Test Yourself

If you think you are already comfortable with this material, try these questions first and mark them using the answers at the back of the book. If you get them all right, you're probably ready to move straight on to the next chapter. If some look tricky, study the chapter first and then come back to these when you're ready.

1. Formally state the well-ordering principle.

2. What is the least element of the set of positive odd numbers?

3. What is the least element of the set $\{x^2 - 8x + 30 | x = 1, 2, 3, \ldots\}$?

4. What is the least element of the set $\{x^2 - 9x + 42 | x = 1, 2, 3, \ldots\}$?

5. What two things must we show in order to prove a rule by induction?

6. Use proof by induction to show $3 + 6 + 9 + \cdots + 3n = \frac{3n^2 + 3n}{2}$.

7. "Guess" and then prove by induction a forumula for:

$$3 + 2 + 1 + 0 + (-1) + \cdots + (4 - n).$$

8. Write the following statement using an "implies" arrow:
   "I always get hot when I drink coffee."

E. Hurst and M. Gould, *Bridging the Gap to University Mathematics*,
DOI: 10.1007/978-1-84800-290-6_12,
© Springer-Verlag London Limited 2009

9. Write the contrapositive to the following statement:
   Diving is permitted here $\Rightarrow$ The water is deep here.

10. Write the following statement using an "implies" arrow, and then write its contrapositive:
    "If x is greater than 7, I know for certain that x is greater than or equal to 7."

12.1 Proof by Induction

Logic lies at the heart of mathematics. One such example of a logical argument, proof by induction, is one of the most widely used tools at the beginning of any numerate degree course. Although to some people induction is a relatively familiar concept, to many it is entirely new. Before we can confidently use induction, however, it is necessary to first formalise "the well-ordering principle."

The Well-Ordering Principle

The natural numbers, $\mathbb{N}$ (i.e., the positive, whole numbers such as $7, 12, 216, \ldots$), have a somewhat obvious yet very important property. If we take any collection of them and then look at the numbers we've taken, we will always be able to identify the *smallest* number.

In the language of logic, which sometimes demands mind-boggling precision, we can write this statement more formally.

- The collection of all the natural numbers is called a set, and so by choosing "some" of them for our collection, we have a *subset* of $\mathbb{N}$.

- We know that our collection must have at least one number in it, so we say formally that it is *"nonempty."*

- Any "member" of a set is called an *element*.

We can tie these facts together to get the formal definition of the well-ordering principle:

"Every nonempty subset of $\mathbb{N}$ has a least element".

Please don't be scared away by how formal this statement looks: mathematicians love writing things like this because there can be no uncertainty as to what is being said – plus there is the added bonus of confusing all those with nonmathematical backgrounds!

EXERCISES

12.1.1. Identify the smallest element of this subset of $\mathbb{N}$:

$$\{1, 3, 7, 8, 9, 11, 13, 15\}$$

12.1.2. Identify the smallest element of this subset of $\mathbb{N}$:

$$\{2n | n = 1, 2, 3, \ldots\}$$

12.1.3. Identify the smallest element of this subset of $\mathbb{N}$:

$$\left\{n^2 - 4n + 6 | n = 1, 2, 3, \ldots\right\}$$

12.1.4. Does every nonempty subset of $\mathbb{N}$ have a *greatest* element? Give a reason for your answer.

12.1.5. Does every nonempty subset of integers (i.e., $\mathbb{Z}$: the positive and negative whole numbers, and 0) have a least element? Give a reason for your answer.

12.2 The Principle of Induction

Now that we have the well-ordering principle under our belts, we'll go for an informal exploration of proof by induction. At the very end of the chapter ("Where Now?"), we'll see how the well-ordering principle is used in the formal statement of this proof.

Imagine we have studied a small amount of data and come up with a "rule" about the data which we *think* might *always* be true. An example could be:

$$1 = 1$$
$$1 + 2 = 3$$
$$1 + 2 + 3 = 6$$
$$1 + 2 + 3 + 4 = 10$$

and "guessing" the rule, "The sum of the first n natural numbers is $\frac{n}{2}(n+1)$."
We could keep trying our "guessed" rule on larger and larger values of n, but
how could we be *absolutely certain* that it would always work, no matter how
large an n we chose? This is where induction comes in.

To start an induction, we always have to show manually that our rule works
for the *first* case. In the example above, the first case is when we have $n = 1$,
and we know that the sum of the first 1 natural numbers is 1. Try this in our
"guessed rule":

$$\tfrac{1}{2}(1+1) = 1$$

Our rule works for $n = 1$.

Now here's the clever bit. Imagine that we could prove this statement:

"In our rule, if it's true for n then it's true for $(n+1)$."

Take a moment to consider the power of this. *If* we could prove it, we could
say that if it's true when $n = 2$ then it's definitely true when $n = 3$. If it's true
when $n = 17893$ then it's definitely true when $n = 17894$.

This is a very good thing, as it reduces the number of times we have to
test our rule in order to be sure that it's valid for lots of different values of n.
Imagine that we wanted to be sure that the rule we "guessed" earlier was true
for $n = 1, 2, 3, 4, 5$ and 6. Before, we would have had to have tested all six cases
separately, but if we knew that

"In our rule, if it's true for n, then it's true for $(n+1)$,"

then testing just $n = 1, 3$ and 5 would automatically ensure that it was valid
for $n = 2, 4$ and 6.

The Lazy Man's Best Friend

Before we go on to what an artist might call the "magic" of induction, you need
to understand the previous bit. Seriously. There is little chance that you'll
be stunned by the power of induction if you don't fully follow what's just
happened, because *things are going to go a little crazy round here.*

We just showed that if we can prove:

"In our rule, if it's true for n, then it's true for $(n+1)$,"

then we can greatly reduce our work in showing that our rule holds true for a certain range of n. By testing just $n = 1, 3$ and 5, we get $n = 2, 4$ and 6 checked "for free." Which is nice. But can we not reduce our work even further?

In checking that the rule works for $n = 1$, we can also be sure that it works for $n = 2$. Now stop. Think for a moment. Do we really need to manually check if the rule works for $n = 3$? *No.*

When we manually check $n = 1$, we are certain that our rule works for $n = 2$. Now, if we look at our original statement:

"In our rule, if it's true for n then it's true for $(n + 1)$,"

We see that once we're sure our rule is true for $n = 2$ then it's definitely true for $n = 3$. Then, if it's true for $n = 3$ it's definitely true for $n = 4$, and so on. See the "magic" at work: we only have to test $n = 1$, and then we get *all* the greater values of n for free. Remember that right at the start of the section we showed:

$$\tfrac{1}{2}(1 + 1) = 1$$

This was our manual check that the rule held for $n = 1$. So if we were somehow able to formally prove:

"In our rule, if it's true for n then it's true for $(n + 1)$,"

then we would know that our rule is always true. This means by testing $n = 1$, we know it works for $n = 2, 3, 4, \ldots, 100, \ldots, 1000000, \ldots$ and so on. Now *that* is a time saver.

Proving the "Magic" Rule

Sadly, with induction, there's no hard-and-fast way of proving our crucial tool:

"In our rule, if it's true for n then it's true for $(n + 1)$."

(That's the last time you have to read it, we promise.) Luckily, though, there is a general strategy that you can follow and adapt. Let's work through our example.

We propose that the sum of the first n natural number is equal to $\frac{n}{2}(n+1)$. We can also write that like this:

$$1 + 2 + 3 + \cdots + n = \tfrac{n}{2}(n + 1)$$

If we add on the *next* (i.e., $(n + 1)$th) term to both sides of this, we get:

$$1 + 2 + 3 + \cdots + n + (n+1) = \frac{n}{2}(n+1) + (n+1) \tag{12.1}$$

(all we have done is added $(n+1)$ to both sides).

Is there some way of rearranging the right-hand side to prove our crucial tool? Let's have a think:

If we say that the sum of the first n natural numbers is $\frac{n}{2}(n+1)$, then what would the sum of the first $n+1$ natural numbers be? It would be $\frac{n+1}{2}((n+1)+1)$, which is $\frac{n+1}{2}(n+2)$. We got this be rewriting my "guessed" rule, but replacing all the ns with $(n+1)$s.

Now take a look at the right-hand side of equation (12.1):

$$\frac{n}{2}(n+1) + (n+1)$$

Manipulate it like this:

$$\begin{aligned}
\frac{n}{2}(n+1) + (n+1) &= \frac{n^2 + n}{2} + n + 1 \\
&= \frac{n^2 + n + 2n + 2}{2} \\
&= \frac{n^2 + 3n + 2}{2} \\
&= \frac{n+1}{2}(n+2).
\end{aligned}$$

We did it. We showed, with our rule, that if it's true for n then it's true for $n+1$, because we can rewrite the rule in $n+1$ and the rule still holds true. As we have already manually tested $n = 1$, we know that our rule works for all n.

Here's a recap and another worked example before you're launched into some exercises:

- There are two "parts" to induction: Part 1 is manually showing that your rule works for the first value of n. Part 2 is showing

  "In our rule, if it's true for n then it's true for $(n+1)$."

  (OK, so maybe the previous time wasn't the last!)

- The general strategy in Part 2 is adding the next term to both sides, then rearranging the right-hand side to prove that your rule works when expressed in $(n+1)$ instead of n.

Here's one more worked example:

Prove that the sum of the first n square numbers is $\frac{n}{6}(n+1)(2n+1)$.

Test $n = 1$:

$$\tfrac{1}{6}(1+1)(2+1) = \tfrac{6}{6} = 1, \text{ which is indeed } 1^2.$$

Our guessed rule can be written:

$$1^2 + 2^2 + \cdots + n^2 = \tfrac{n}{6}(n+1)(2n+1)$$

Add on the next term:

$$1^2 + 2^2 + \cdots + n^2 + (n+1)^2 = \tfrac{n}{6}(n+1)(2n+1) + (n+1)^2$$

Rearrange the right-hand side:

$$
\begin{aligned}
\frac{n}{6}(n+1)(2n+1) + (n+1)^2 &= (n+1)\left(\frac{n}{6}(2n+1) + n + 1\right) \\
&= (n+1)\left(\frac{2n^2 + n}{6} + n + 1\right) \\
&= (n+1)\left(\frac{2n^2 + 7n + 6}{6}\right) \\
&= \frac{(n+1)(2n+3)(n+2)}{6} \\
&= \frac{n+1}{6}(n+2)(2n+3) \\
&= \frac{n+1}{6}((n+1)+1)(2(n+1)+1).
\end{aligned}
$$

Done!

EXERCISES

12.2.1. I look at this:

$$1 = 1$$
$$1 + 1 = 2$$
$$1 + 1 + 1 = 3$$

and I guess the rule $\underbrace{1 + 1 + 1 + \ldots}_{n \ times} = n$. Prove my rule by induction.

12.2.2. I look at this:

$$2 = 2$$
$$2 + 4 = 6$$
$$2 + 4 + 6 = 12$$

and I guess the rule $2 + 4 + 6 + \cdots + 2n = n^2 + n$. Prove my rule by induction.

12.2.3. Prove the following statement by induction:

$$1^3 + 2^3 + 3^3 + \cdots + n^3 = (1 + 2 + 3 + \cdots + n)^2$$

(Remember that $1 + 2 + 3 + \cdots + n = \frac{n}{2}(n + 1)$).

12.2.4. "Guess" and then prove by induction a formula for the sum of the first n odd numbers.

12.2.5. The set m is defined as:

$$m = \{4k \,|\, k = 1, 2, 3, \ldots\}$$

Find and prove by induction a formula for the sum of the first n elements of m.

12.3 Contrapositive Statements

Implications

At this step we're going to need a new symbol. It looks like this: $\Rightarrow$, and it means "implies." Here are some examples of it in action:

- I have exactly two American coins in my pocket $\Rightarrow$ I have at least 2 cents in my pocket.

- I wear strong reading glasses $\Rightarrow$ I do not have good eyesight.

- It is 3 pm $\Rightarrow$ It is after midday.

The first and most useful thing to note is the direction of the arrows: "implies" only goes one way. There is a different symbol ($\Leftrightarrow$) for statements where the "implies" works both ways.

Let's just clarify things. Consider again the statement:

It is 3 pm $\Rightarrow$ It is after midday.

If we say "it is 3 pm," we immediately know *for certain* that it is after midday. But consider the implication the wrong way around: if we say "it is after midday" this in no way implies that it is 3 pm. It could be 12:01 pm and our statement that "it is after midday" is still certainly true, but it is most certainly *not* 3 pm at this time! This special "feature" of the implies arrow is what gives us the ability to deduce contrapositive statements.

It's All Negative

As we've just seen, implications do *not* work backwards. But if we *negate* the statements we can do a clever trick. Negating a statment means changing "true" to "false" in the statement. For example:

Statement	Negated Statement
It is 3 pm.	It is not 3 pm.
I do not have good eyesight.	I have good eyesight.
I have at least 2 cents in my pocket.	I have less than 2 cents in my pocket.

(consider that last one carefully — writing it as an inequality is helpful).

We form a contrapositive statement in two steps:

- Negate both sides of the implication.

- Reverse the arrow.

So the contrapositives to our original three statements become:

I have less than 2 cents in my pocket. $\Rightarrow$ I do not have 2 coins in my pocket.

I have good eyesight. $\Rightarrow$ I don't wear strong reading glasses.

It is not after midday. $\Rightarrow$ It is not 3 pm.

Hopefully this is sufficient exploration for you to tackle some exercises. Remember: negate both sides and reverse the implication. Easy!

EXERCISES

12.3.1. Write the contrapositive to the following statements:

- I have a phobia of heights $\Rightarrow$ I have never been up the Eiffel Tower.

- x is a prime number greater than 2 $\Rightarrow$ x is odd.

- The sum of the digits of a positive whole number is divisible by 3 $\Rightarrow$ The number itself is divisible by 3.

- x is positive $\Rightarrow$ $x > 0$

12.3.2. The definition of a prime number is "a positive integer, exactly divisible by *only* 1 and itself." How can we tie together two of the above statements to help us part of the way to determining whether an integer chosen at random is prime?

Where Now?

There are many other proof methods besides those outlined in this chapter. Another common method of proving something is *proof by contradiction*. This method can prove an implication by showing that if the implication *weren't* to be true, we must arrive at a contradiction. For example, if we were asked to prove the statement, "All positive integer multiples of 4 are even," we could proceed as follows:

1. Assume that there exists n, a positive integer multiple of 4 that *isn't* even.

2. Then $n = 4z$, where $z \in \mathbb{Z}$.

3. So $n = 2 \cdot (2z)$, $z \in \mathbb{Z}$.

4. So n is an even number, which is a contradiction to the original assumption that "n is not even."

5. So there cannot possibly exist such an n.

6. So all positive integer multiple of 4 must be even.

Hopefully you'll agree that this method of proof is neat and very useful!

The next logical thing is to formalise our rather informal statement of induction. This follows from the well-ordering principle, and the idea of the proof is to formulate a contradiction. We start by assuming that there *are* some values of n that our rule won't hold for. By the well-ordering principle, of these numbers there must be a least element. But we manually checked the *first* possible value of n, so we know it's true for the value of n directly before this least element, and by our "crucial tool" we know that it *can't* be false for the next value of n, thereby forming a contradiction. There are therefore no finite values of n that the rule won't hold for.

Taking the time to become comfortable with induction is most certainly time well spent. *What Is Mathematics? 2nd Ed.* (R. Courant, H. Robbins and I. Stewart, Oxford University Press, 1996) is a great place to look: there's a detailed explanation of the process and some excellent examples to practice with too. If you enjoyed exploring the logic behind induction you might also like to learn about "strong induction," which will be another useful tool at university.

Contrapositive statements are very, very helpful in analysis, because often statements are useful to us in both their original and contrapositive forms. This way, if we know a function $f(x)$ *never* behaves in a certain way, if we note this behaviour in some unknown we're looking at, we know for certain that we are not looking at function $f(x)$.

Thinking back to the chapter on sets and functions, it is very helpful to write the definition of injectivity in symbols. Our definition, in words, was:

> If we have a function, and we also know that we have only one value of x for each value of y, then our function is *injective*.

In symbols, this becomes:

> If $f(a) = f(b) \Rightarrow a = b$, then f is injective.

The reason that we want to put the definition into symbols is so that we can take the *contrapositive* of this statement, which gives us the equally useful:

> If $a \neq b \Rightarrow f(a) \neq f(b)$, then f is injective.

It's crucial to remember that *both* of the above statements are a correct definition of injectivity, and both versions are very helpful to us. By taking the contrapositive we aren't defining anything new, nor are we setting out conditions for functions *not* to be injective: we are finding an *equivalent* definition for the same idea.

13
Probability

Test Yourself

If you think you are already comfortable with this material, try these questions first and mark them using the answers at the back of the book. If you get them all right, you're probably ready to move straight on to the next chapter. If some look tricky, study the chapter first and then come back to these when you're ready.

1. Find the value of 9 factorial.

2. State the value of 0!

3. How many ways are there of choosing four cards from a set of ten cards?

4. State, using factorials, the definition of $\binom{n}{r}$.

5. Twenty-five people stand outside a television studio. In a moment, 20 computers will each pick one person, independently and at random. If chosen, a person will be allowed into the studio to be in a programme's audience. How many people should expect *not* to get in?

6. In a casino, you see a new game on offer. The dealer deals you ten cards at random from a single deck of cards. You win if you are dealt one heart, two clubs, three diamonds and four spades, but you lose if you are dealt anything else. What is the probability that you win this game?

E. Hurst and M. Gould, *Bridging the Gap to University Mathematics*,
DOI: 10.1007/978-1-84800-290-6_13,
© Springer-Verlag London Limited 2009

7. 20% of new mobile phones are produced in Wales, and the other 80% are produced in Scotland. The probability of a mobile phone from Wales being faulty is found to be 10%, and the probability of a mobile phone from Scotland being faulty is found to be 5%. You buy one of these new mobile phones from the shop, and are pleased to find that it is not faulty. What was the probability of this happening?

8. What is the probability of rolling a 4 on a fair, six-sided die, given that you roll at least a 4?

9. There are 100 tiles in a bag. There are two of every tile, and the tiles show the numbers 1 – 50 (i.e. there are two 1s, two 2s, two 3s...). Given that you have drawn a number that is greater than 45, what is the probability that the next tile you draw will show a number greater than 40?

10. In a town, new cars are only sold in two showrooms. The showroom on the north side of the town sells 30% of the cars, and the showroom on the south side of the town sells 70% of the cars. 20% of cars sold from the north showroom are red, and 40% of cars sold from the south showroom are red. Given that the last car sold in the town was *not* red, what is the probability that it was sold on the north side of town?

13.1 Turn that Frown Upside-Down

Before university, probability is always grouped together with statistics. While many students find this topic interesting and exciting, there are plenty that don't, and as such many people overlook just how useful probability is. Here, we're going to look at probability as a topic in its own right, not just as a part of the world of statistics. In this chapter we're going to first look at the basics of probability, and then move on to using various "tricks" to solve problems using probability. Finally, we'll take a look at the idea of conditional probability, and see how that can be a useful tool in solving many problems – even ones that we encounter in everyday life.

In the Beginning...

Before we can start working with probability problems, we first need to explore some of the tools that we're going to be using to solve them. Some you'll have met before, but some you may not have, so make sure that you're confident with all of these concepts – *and* why they're useful to us. Here we go:

Factorial: "Factorial" is quite a common weapon in solving probability problems. Quite simply, to take "the factorial" of any natural number n, we simply find the value of $n \cdot (n-1) \cdot (n-2) \cdots 2 \cdot 1$. That is, we multiply the number by itself minus 1, then itself minus 2, then itself minus 3,... all the way down to multiplying it by 1. To denote "factorial," we put an exclamation mark after the number that we are finding the factorial of. Here are some examples:

- $2! = 2 \times 1 = 2$.

- $4! = 4 \times 3 \times 2 \times 1 = 24$.

- $6! = 6 \times 5 \times 4 \times 3 \times 2 \times 1 = 720$.

One thing to note about factorials is that $0! = 1$. Although this doesn't follow the way that we've defined things above, a whole host of problems that we solve using factorials go horribly wrong if we don't insist that $0! = 1$, so don't forget it! Factorials aren't defined for negative numbers, nor are they for positive numbers that aren't whole numbers.

Choose: The idea of "choosing" is one that will hopefully appeal to your logical intuition. In probability problems, we often need to ask the question, "In how many different ways could I choose r balls from the n possible balls?" Initially, you might think the use of this is rather limited, but in reality this is one of the most widely used ideas in probability problems. We would need to use it in situations like these:

- You are dealt five cards from a full deck of 52. What is the probability that you are dealt exactly three spades?

- You roll a fair, six sided die ten times. What is the probability that you roll five 1s?

- You flip a coin ten times. What is the probability that you get exactly eight heads?

The difficulty of these problems varies quite a lot, but the idea of "choosing" is crucial in all of them. Let's look at the middle example about rolling a die. We know that the probability of rolling a 1 is $\frac{1}{6}$, and so the probability of *not* rolling a 1 is $\frac{5}{6}$. If you're going to roll the die ten times, you're going to need five 1s and five "not 1s." But we can't just multiply all these ten fractions together; we also need to take into account the fact that you might roll your ones on the first five rolls, on rolls $2, 4, 6, 8$ and 10, on rolls $1, 4, 5, 7$ and 10, ..., there are many, many possibilities. How many exactly? 10 choose 5.

So what exactly is this elusive "choose" function? Well, we write "n choose r" in one of two ways, and its definition is as follows:

$$^nC_r = \binom{n}{r} = \frac{n!}{r!(n-r)!}$$

A few clever tricks to do with the factorial function clarify exactly where this odd looking expression comes from. Expanding both the factorials in n allows for a cancellation of the last $(n-r)$ terms:

$$\frac{n!}{r!(n-r)!} = \frac{n \cdot (n-1) \cdot (n-2) \cdots 2 \cdot 1}{r!(n-r) \cdot (n-r-1) \cdot (n-r-2) \cdots 2 \cdot 1}$$
$$= \frac{n \cdot (n-1) \cdot (n-2) \cdots (n-r+1)}{r!}$$

To then see where the numerator comes from, think of a situation in which you might want to put r balls into n boxes. Once a box has a ball in it, it can't fit another ball in too. Therefore there are n choices of box in which to put the first ball, $n-1$ choices of box in which to put the second ball, $n-2$ choices of box in which to put the third ball, and so on until there are $n-r+1$ choices of box in which to put the rth ball. Finally, if we assume that the balls are all identical, it's not possible to know which *order* we put the balls into the boxes. Because we put r balls into boxes altogether, there are $r!$ different orders that we could have placed the balls in the boxes (because there were r balls to choose as the "first" ball, $r-1$ to choose as the "second" ball and so on), so in finding the binomial coefficient it is necessary to divide by $r!$ to take into account the fact that we *don't* know what order the events occurred in.

Independent Events

One final concept that we're going to need later is that of independent events. We say that two events are independent if the probability of one of them happening has no impact on the probability of the other happening. When we're working with probability problems, it's crucial that we work out as soon as possible whether events are independent. For example, to find the probability of rolling a 4 on a fair, six-sided die *and* flipping tails on a fair coin, we simply do $\frac{1}{6} \times \frac{1}{2} = \frac{1}{12}$, because the events are independent. However, to find the probability of success in the task "Draw a yellow ball then a green ball from a bag of two yellow and two green balls, without replacement between the draws" we need to spot straight away that the events are not independent: if we succeed in drawing the yellow ball first, the probability of then drawing a green ball is $\frac{2}{3}$, because we don't replace the yellow ball, whereas the probability of drawing a green ball if no other ball had been drawn is $\frac{1}{2}$. The moral of the story here is to decide whether your events are independent *before* you run headfirst into the problem. Now for some exercises:

EXERCISES

13.1.1. What is the value of 5 factorial?

13.1.2. Find 8!

13.1.3. How many ways are there of choosing three balls from a set of five balls?

13.1.4. Find the value of $^6C_2$.

13.1.5. Find the value of $\binom{8}{7}$.

13.1.6. Using factorials, write the definition of "n choose r."

13.1.7. What is the probability of rolling "3 or 4" on a six-sided die?

13.1.8. What is the probability of not drawing an orange ball from a bag of four grey, six purple and ten orange balls?

13.1.9. Are the events involved in, "Draw the 7 of hearts and then draw the 3 of diamonds, at random from a standard deck of 52 cards" independent if we *do* replace the card that is drawn first before drawing the second?

13.1.10. In finding the probability of the event "Draw the 3 of clubs, then draw the 2 of clubs, at random from a standard deck of 52 cards," are the two individual events independent if we *do not* replace the card that is drawn first before drawing the second?

13.2 Solving Probability Problems

In this section we're going to look at actually *solving* some problems to do with probability. Rather than going through lots and lots of text before actually tackling the problems, we're going to jump straight into them and explain the logic of our steps as we go. We're also going to look into the pitfalls along the way – learn from the potential mistakes so that you don't make them!

Clay Pigeon Shooting

Ten hunters enter a field. In a moment, ten clay pigeons will be launched into the air. Every hunter will simultaneously fire just one shot and they will each *definitely* hit whichever clay pigeon they have randomly chosen to shoot at. Find out how many clay pigeons the hunters should expect to survive.

First of all, if you're tutting, thinking of answering 0 and heading off to the next question, *be ashamed!* You've not grasped the subtlety of the question. Each hunter is going to choose the clay pigeon that they fire at *at random*, and so it's quite possible that all hunters will shoot at the *same* clay pigeon, and then there would be nine that survive. Let's think about this mathematically: Each clay pigeon has a $\frac{1}{10}$ chance of being shot by any particular hunter. But is this particularly helpful to us? We would still need to list all of the possible combinations of which hunter shoots which bird. Then they would all be equally probable, and only *then* would we be able to make any use of this fact.

Instead, we're going to use the idea of "not": when we look at the probability of something *not* happening, we simply find 1 minus the probability that it does occur. So the probability of a given bird *not* being hit by a certain hunter is $\frac{9}{10}$. Because the hunters choose their targets at random, we can say that the probability of a given bird not being hit by *any* of the hunters is $\left(\frac{9}{10}\right)^{10}$. This comes from the fact that the events, "Hunter x doesn't hit this bird," and, "Hunter y doesn't hit this bird," are *independent* for all pairs of x and y: remember, targets are chosen at random.

So, if the probability of a given bird surviving is $\left(\frac{9}{10}\right)^{10}$, and all of the birds are equally likely to survive, then to find the number of birds that we should expect to survive we just multiply the chance of a bird surviving by the number of birds. Our answer, then, is $\left(\frac{9}{10}\right)^{10} \times 10 = 3.49$. Because only a whole number of birds can survive, the hunters should expect that three birds survive.

Cards Games

You are playing a new card game. In this game, you choose five cards at random from a deck of 52 standard playing cards. You win if you choose three black cards and two red cards. What is the probability of winning?

First of all, let's look at the common mistake: rushing in with the idea, "The chance of drawing a red card is $\frac{1}{2}$ and the chance of drawing a black card is $\frac{1}{2}$, so the answer is just $\left(\frac{1}{2}\right)^5$. This is *wrong*. It neglects to spot the fact that the events *are not* independent. We're going to need to use much better logic to tackle this one.

Firstly, let's think about which of the tools from the previous section we're going to need here. We have to have three black and two red cards as our five, but the order doesn't matter. Sounds like a "choosing" problem.

Now, let's think about the problem like this: we need to find the number of winning possibilities, and then divide this by the total number of possibilities. This will give us the probability that we need.

Firstly, the denominator (the total number of possibilities) is far easier to find. This is just "the number of ways of choosing five cards from 52" – this is simply $\binom{52}{5}$.

Now for the numerator – the number of ways of winning. To win we need to have drawn three cards from the 26 black ones and two cards from the 26 red ones. Again, we have a choosing problem! The answer to this one is just $\binom{26}{3} \times \binom{26}{2}$: we *multiply* these together because we need *both* events to happen. We now have everything we need for our answer:

$$\frac{\binom{26}{3} \times \binom{26}{2}}{\binom{52}{5}}$$

We can simplify this and give a decimal as our answer or leave it in this form if we don't have access to a calculator.

Cracked Glasses

A new glassware company produces vases in two locations: 30% of their vases are made in London, and 70% of their vases are made in New York. Of the vases that are produced in London, 5% are faulty. Of the vases that are produced in New York, 10% are faulty. What is the probability that a vase chosen at random is faulty?

The obvious mistake to make here is to try to blindly "average" the two percentages of faulty vases. While we are going to do an averaging of sorts, what we're actually aiming for is a *weighted* average because the probability of a vase coming from London is not the same as the probability of that vase coming from New York.

We're definitely going to need a multiplication here, but what are we actually aiming for? Let's think back to "and" and "or" from Chapter 11. What statement do we need to make to ensure that we'll catch all of the faulty vases? How about this:

(Faulty and from London) or (Faulty and from New York)

That looks pretty good to me – it makes sure that we're going to look at all of the faulty vases, and it takes into account the origin of a vase before trying to assign a probability. Now, when we see the word "and," in probability we multiply because we need *both* of the statements to be true. When we see the word "or" and we're dealing with mutually exclusive events (this means that if one happens, the other *definitely doesn't*, and vice versa) in probability we add, because we're happy for either of the events to occur. So we get the final answer as being $(0.05 \times 0.3) + (0.1 \times 0.7) = 0.085$. This seems reasonable, as it lies between the two probabilities of being faulty that are dependent on location.

Right, here are some similar problems for you to tackle yourself. Take your time and make sure that you are *logical* in your approach.

EXERCISES

13.2.1. There are 20 candidates in a game show. Each of 20 computers will simultaneously pick, independently and at random, one of the candidates to be eliminated. How many candidates should expect to survive this first round of eliminations?

13.2.2. In the game show, the computers reveal their selections and ten candidates make it through the first round of eliminations. Now the 20 computers each pick one of the ten to be eliminated. This choice is again made independently and at random. How many candidates should expect to make it through this time?

13.2.3. When rolling a fair, six-sided die ten times, what is the probability of rolling a 2 on the first eight rolls and then a 6 on both the 9th and 10th rolls?

13.2.4. When rolling a fair, six-sided die ten times, what is the probability of rolling 2 eight times and a 6 twice?

13.2.5. A newspaper gives away free CDs as part of a promotion. Seven out of every ten of their papers contains a "Pop Hits" CD and three out of every ten of their papers contains a "Rock Hits" CD. Because it is in a sealed bag, people cannot see which CD they're getting until after they have bought the paper. The newspaper thinks that there is a 90% chance that a person chosen at random would enjoy the "Pop Hits" CD, and a 30% chance that a person chosen at random would enjoy the "Rock Hits" CD. You speak to a friend who bought the paper this morning. What is the probability that your friend enjoyed the CD?

13.2.6. A new ornament has been released by a manufacturer. They made 1000 of the ornaments in total, but 350 of them were made from glass and the rest were made from clay. The ornaments are sold in a sealed box, so when buying the ornament it is not possible to tell whether you have a glass or a clay edition. Upon an impact, the glass ornaments have a $\frac{9}{10}$ probability of breaking, and the clay ornaments have a $\frac{2}{10}$ probability of breaking. To resell, a broken ornament is worth nothing regardless of what it is made of, a glass ornament is worth £100 and a clay ornament is worth £10. You just bought an ornament, but then dropped it before you could see what it was made of. What is the expected value of your ornament now?

13.3 Conditioning

There's one final area of elementary probability left to explore: conditional probability. We can use conditioning in a massive variety of situations, but it's one of those things that's a bit strange to learn. You see, while the previous section appealed to your "instinct" in deciding how to proceed, conditioning is not so pleasant. While you're learning the ropes, you might even feel that what you're doing is sort of *wrong* – but as long as you follow all of the steps carefully, you'll be fine.

Given That...

The idea of conditioning is to find the probability of something happening, *given that something else has already occurred*. Let's think about the weather, for example: If we were trying to guess the probability that it is going to be sunny tomorrow, we might guess $\frac{3}{10}$. But how about if we also knew that it is sunny today? Now, we might guess $\frac{4}{10}$: that is, we are saying that the probability that it is sunny tomorrow, *given that it is sunny today*, is $\frac{4}{10}$.

This sort of example is fine for illustrating the *gist* of conditional probability, but we can make much more precise use of the idea. Imagine that we have just rolled a fair, six-sided die, and that we haven't seen the result of the roll but someone else has. What is the probability that we have just rolled a 3? $\frac{1}{6}$, I hear you cry. But what if the person who has seen the result of the roll says that the number rolled was odd? Now what is the probability that we have just rolled a 3? $\frac{1}{3}$: we know that the only possible outcomes are 1, 3 and 5 now, so our odds are much better. Written in the language of conditioning, the probability of rolling a 3 is $\frac{1}{6}$, but the probability of rolling a 3 *given that* we have rolled an odd number is $\frac{1}{3}$.

We can write all of these ideas more formally. When we say, "The probability of A given B", we write $P(A|B)$. Make sure that you remember which order the A and B are in that expression, because the probability of A given B is most certainly *not* the same as the probability of B given A.

In making our calculations, if we have a situation that is more complicated than the one described above, it's very helpful to have a formula to work with:

$$P(A|B) = \frac{P(A \cap B)}{P(B)}$$

In words, the right-hand side of this expression is, "The probability of both A and B occurring, divided by the probability of just B occurring." Let's consider why it's logical that this is indeed the definition: if we want to test whether A occurs *given* that B has just occurred, by the end of our "test" for A to have occurred then *both* A and B will need to have occurred (because B already has). But we *know* that B has already occurred, and so in finding the probability of A given B, we need to divide the probability of both A and B occurring by the probability of B occurring. This is the one and only definition in this section before we move on to examples, so learn it well!

> A spinner is divided into two halves, left and right. The left half contains the numbers 1 - 8 and the right half contains the numbers 9 - 16. All numbers are equally likely to occur. What is the probability that you score at least 7 on the spinner, given that you get a score from the left half?

For those of you that can visualise this problem and "see" that the answer is $\frac{1}{4}$, please feel free to head to the next example. If you'd rather work with the formula, we need to find two things:

- The numerator of the fraction is $P(A \cap B)$. This is the probability of getting at least 7 *and* getting a score from the left half, which is altogether equal to $\frac{2}{16}$. This is because in the left half, only the numbers 7 and 8 are good for us, so these are just two numbers out of the possible 16.

- The denominator of the fraction is $P(B)$. This is the probability of getting a score from the left half, which is simply $\frac{1}{2}$, because the spinner is divided in half.

So the solution is just:

$$\frac{P(A \cap B)}{P(B)} = \frac{\frac{2}{16}}{\frac{1}{2}} = \frac{1}{4}$$

One final example to end the chapter, then some exercises:

Your friend draws a card at random from a pack of 52 cards and shows you that he has drawn a club. Then, without your friend replacing his card, you draw a card independently and at random from the remaining 51 cards. What is the probability that the card you drew is a club?

We'll tackle this question from both a logical and conditioning angle. Firstly, simply thinking carefully about the problem is a perfectly valid way of obtaining your solution. Your friend shows you that he drew a club, so of the 51 cards remaining there are 12 clubs. Therefore the probability that you're holding a club is $\frac{12}{51} = \frac{4}{17}$).

Alternatively we can turn to the conditioning formula. The probability of you drawing a club *and* him drawing a club is the probability of drawing two clubs from the deck. We know from earlier that this is a choosing problem, so the answer is:

$$\frac{\binom{13}{2}}{\binom{52}{2}} = \frac{1}{17}$$

Now, to find the denominator of our conditioning equation we simply find the probability that your friend drew a club. Easy – it's just $\frac{1}{4}$. Now we just do the final division to get our answer:

$$\frac{\frac{1}{17}}{\frac{1}{4}} = \frac{4}{17}$$

There we have it – exactly the same answer from two entirely different approaches.

EXERCISES

13.3.1. What is the probability of rolling a 6 on a fair, six-sided die *given that* you rolled at least a 3?

13.3.2. What is the probability of rolling a 2 on a fair, six-sided die *given that* you rolled an odd number?

13.3.3. You make a spinner that has four equal sections: blue, red, yellow and green. What is the probability you spin green, given that you don't spin yellow?

13.3.4. You pick a card at random for a set of cards labelled 1 - 100. Given that you have chosen a number that displays a multiple of 12, what is the probability that you have chosen the number 60?

13.3.5. Two factories produce CDs. 40% of CDs are produced in Newcastle and 60% of CDs are produced in Manchester. In the Newcastle factory 5% of the CDs produced are faulty and in the Manchester factory 10% of the CDs produced are faulty. Given that a CD is faulty, what is the probability that it was produced in Newcastle?

Where Now?

The tools we've looked at here are the building blocks to some very exciting ideas in probability theory. Hopefully you've seen that these ideas are rigorous and relevant – two of the main things that many students claim are lacking from statistics before university.

From here, there are some very famous problems that can be tackled by adapting the skills covered in the chapter. For example, if you flip a coin ten times in a row, how would you find out the probability of getting exactly 4 "runs" of heads and tails? It sounds tricky, but it uses something called "occupancy theory", which is really only a minor extension to the choosing problems we explored above.

Imagine that the government need to test for a new blood disease that affects 0.01% of the population. They could save a lot of time and money by "pooling blood." They do this by mixing n samples of blood together and only doing the test once on this mixture. If the whole sample is disease-free, everyone in the sample is clear, but if the sample is not disease free, everyone in the sample needs to be tested individually to see exactly who is infected. The question is, what is the optimum value of n? Grouping too few people together will mean that the government has to do a huge number of tests in order to test everyone. Grouping too many people together is also foolish because then many of the mixed samples will test positive for having the disease, and then all of the people involved in these samples need to be tested individually. The way to solve such a problem starts just like the "clay pigeon shooting" example, where we need to look at "have nots" rather than "haves", but from there we move on to calculus in order to optimise our choice. I'm sure you'll agree that these problems are far more involved than you might have expected!

Probability (J. Pitman, Springer-Verlag, 1993) offers a fantastic insight into many of the exciting uses of statistics. If you want to brush up on your skills (and even learn some new ones) before starting your degree, we'd strongly recommend that you take a look. There's a lot of difficult material there, so you'll need to dip in and out, but you'll definitely be well ahead of the pack if you take the time to look at even just a couple of the sections.

14
Distributions

Test Yourself

If you think you are already comfortable with this material, try these questions first and mark them using the answers at the back of the book. If you get them all right, you're probably ready to move straight on to the next chapter. If some look tricky, study the chapter first and then come back to these when you're ready.

1. Which probability model would we use when we are looking at a series of events that occur at a known average rate through time?

2. Which probability model would we use when we are looking at a series of independent repetitions of the same trial, with fixed probability of success?

3. A card is drawn at random from a standard deck of 52 playing cards. The number on the card is noted and the card is replaced. This is repeated until ten cards have been noted. What is the probability that exactly four aces were drawn?

4. A ball is drawn at random from a bag containing two yellow, three red and four green balls. The colour is noted and the ball is replaced. What is the probability of drawing two yellow balls if the process is undertaken five times?

5. A London Underground train driver notes that over the course of a journey, the average number of people boarding the train is 250. What is the

E. Hurst and M. Gould, *Bridging the Gap to University Mathematics*,
DOI: 10.1007/978-1-84800-290-6_14,
© Springer-Verlag London Limited 2009

probability that, on a given journey, *exactly* 250 people board the train?

6. A cleaner at an art gallery has to clean around all 1000 exhibits each day. They note that 1% of exhibits have litter left on them in the course of a day. What is the probability that five exhibits will have litter left on them today?

7. If $X \sim B(10, 0.2)$, what is the mean and variance?

8. If $X \sim Po(5)$, what is the mean and variance?

9. An astronaut training programme has 5000 students. The probability that any given pupil will go on to visit the International Space Station is 0.001. What is the probability that six of these students go on to visit the International Space Station?

10. A mechanic has two new car batteries in stock. Of the 1000 people that visit his shop each day, the mechanic notes that on average, the probability of a person requiring a new car battery is 0.003. What is the probability that he won't be able to meet the demand for new batteries today?

14.1 Binomial Events

One of the key ideas in the world of probability is that of a Bernoulli trial. A Bernoulli trial is any single event with only two possible outcomes (success or failure) and with a fixed probability of success. An example of a Bernoulli trial is "rolling a one on a fair, six-sided die," with the fixed probability of success being $\frac{1}{6}$.

This is all very well, but how is this useful to us on a larger scale? Well, imagine that we were going to roll that same die 100 times. What is the probability that we roll exactly three 1s? Or exactly 99 1s? This is where we need the idea of *binomial distribution*.

The binomial distribution gets its name from the fact that it uses the same basic principle as binomial expansion of brackets. We're going to be using the idea of "choosing," which was covered in the previous chapter, so if you're unfamiliar with this you'll need to review it now.

Let's think again about the example with the die. If we're trying to find the probability that we roll exactly three 1s out of a hundred rolls, then we're going to need three "successes" and 97 "failures." The probability of any given trial being a success is fixed at $\frac{1}{6}$, and the probability of any given trial being a failure is fixed at $\frac{5}{6}$. All of these events are most certainly *independent* (this term was covered in the previous chapter) and so we need to multiply things together to get our desired answer. That means we're going to multiply three lots of $\frac{1}{6}$ and 97 lots of $\frac{5}{6}$ together, giving us $(\frac{1}{6})^3(\frac{5}{6})^{97}$. But wait: what about the arrangement of these events? Hopefully you've guessed it: we need to take into account the number of way that there are of choosing our three successes from the 100 trials. This will give us the coefficient $\binom{100}{3}$ (i.e., 100 choose 3) that we also need to multiply by, giving us the final expression:

$$\binom{100}{3} \left(\frac{1}{6}\right)^3 \left(\frac{5}{6}\right)^{97}$$

The Wider World

Now that we've covered a specific example, let's attempt to modify what we did so that we could apply it to *any* situation where we would use binomial distribution:

- Consider just a single trial. Find the probability of this single trial being a success.

- Determine how many trials there are to be altogether.

- Decide how many successes you want to check for.

- Where n is the number of trials, p is the probability of success, q is the probability of failure (and, as such, $q = 1 - p$) and r is the number of successes that you are checking for, substitute everything into the following equation:

$$P(X = r) = \binom{n}{r} p^r q^{n-r}$$

Are you happy with where all of those steps come from? If not, review the example with the die and compare what we're doing there with what we do in the general case.

There is a standard notation that we use when we're dealing with the binomial distribution. If we wanted to say, "There is a binomial event in which we are going to find the probability of X successes from a total of ten trials, all with success probability $\frac{3}{20}$," then we write $X \sim B(10, \frac{3}{20})$. That is, in general, for n trials with success probability p, we write $X \sim B(n, p)$, and we read the symbol $\sim$ as "is distributed."

Now for some exercises:

EXERCISES

14.1.1. What is the name of a single trial with only two possible outcomes and a fixed probability of success?

14.1.2. What is the name of the distribution that we use when we have many of these trials independently and wish to find the probability of achieving r successes?

14.1.3. What is the probability that in flipping a coin ten times you get ten heads?

14.1.4. What is the probability that in flipping a coin 20 times you get ten heads?

14.1.5. I choose a card at random from an ordinary deck of 52 playing cards, note the suit of the card and then replace it. Altogether, I undertake this procedure 25 times. What is the probability that I will have noted less than two hearts in this time?

14.1.6. I play a game in which I randomly choose a ball from a bag. The bag initially contains ten balls, five of which are green and five of which are orange. I then repeat this process four more times, drawing from the same bag without replacing the balls that I have chosen each time, giving me a total of five balls chosen. I win the game if I have chosen exactly four orange balls. Why is it not appropriate to model this event under a binomial distribution?

14.2 Poisson Events

The Poisson distribution is similar to the binomial distribution in that it can easily be used simply by remembering a formula. The problem with the Poisson distribution is making sure that the conditions are right so that it is definitely appropriate to do so.

We use the Poisson distribution to model events that happen at a fixed rate in an interval, and that interval is often time. A good example of this is a receptionist at the switch board of a company receiving telephone calls. From years of experience, she knows that in a 10 minute block of time she will receive, on average, 20 phone calls. We can use the Poisson distribution to find the probability that she receives two phone calls, or that she receives 40 phone calls, in any given 10-minute block. The Poisson is appropriate here because we're looking at events that occur randomly in time, but with a known "average rate."

Another example of when we could use the Poisson distribution is this: A machine makes 100 curtains per hour. It is known that 2% of the curtains that the machine makes are faulty. What is the probability that, in a given hour, the machine makes ten faulty curtains?

Both of these questions are real-life examples of where we can put the Poisson distribution to work. The company that employ the receptionist might use the Poisson distribution to find out if it is likely that there will be times that the receptionist has too many calls to efficiently handle, and the curtain manufacturers may need to know the probability that there are times that the machine is making so many faulty products that it would be cheaper to replace the machine than keep destroying faulty curtains. Firstly, let's formalise the list of requirements that we have to satisfy in order to actually *use* the Poisson distribution, and then we'll go about solving these two problems.

In order to be able to model an event using the Poisson distribution, we need to ensure:

- Every event is independent.

- The average rate does not change over the interval we examine.

- The number of events that occur in a given time period depends only on the length of the time period and the average rate of occurrances.

- Two or more events cannot occur at exactly the same instant.

Just like the binomial distribution, there is a standard way of writing down events that follow the Poisson distribution. Because now we only have one parameter to deal with (the fixed rate that the event occurs at), we simply write $X \sim Po(\lambda)$, where λ is this fixed rate.

Now on to solving the two problems set out. We're going to need a formula like we did for the binomial distribution, but this time we're just going to *use* it, not look too deeply into why it works. You'll cover the logic behind why the formula is what it is at university, but being able to use it is definitely a big step in the right direction. So, stated without proof, here is the important formula that needs to be committed to memory:

$$P(X = r) = \frac{e^{-\lambda}\lambda^r}{r!}$$

Remember that the $r!$ on the bottom line means "r factorial" – this was covered in the previous chapter. Right, back to the problems:

With the receptionist, we know that the average number of incoming calls in any given 10-minute block of time is 20. This means that we can write

$X \sim Po(20)$. To find the probability of there being only two phone calls in 10 minutes, we use have the value $r = 2$. All that remains is a big substitution:

$$P(X = 2) = \frac{e^{-20} \cdot 20^2}{2!}$$
$$= 200e^{-20}$$

If we go to a calculator we can get the decimal answer (0.000000412, to 3 s.f.), but in reality leaving our answer in terms of e is perfectly acceptable (many universities do not allow calculators in exams whatsoever, so you might need to get used to not using them anymore!). Now for the receptionist's second scenario, where we need to find the probability that she receives 40 calls in a given 10 minutes. This time we're looking at:

$$P(X = 40) = \frac{e^{-20} \cdot 20^{40}}{40!}$$

Because numbers like 20^{40} and 40! are so enormous, there's not really any simplification we can do here – it's perfectly OK to leave the answer like this if you're without a calculator (for those of you desperate to use one, it's 0.0000278 to 3 s.f.).

Finally, let's take a look at our curtain making machine. This time, we have a tiny extra step to do: we need to actually calculate the value of λ, because we're not given it explicitly. What we *are* told is that, on average, 2% of the curtains are faulty, and that it produces 100 curtains per hour. This means we can quickly see that, for any given hour $\lambda = 2$, because this is 2% of the 100 curtains that it produces.

All that's left is using the equation again. We need to find the probability that the machine makes ten faulty curtains in an hour, so that's going to be:

$$P(X = 10) = \frac{e^{-2} \cdot 2^{10}}{10!}$$
$$= \frac{1024e^{-2}}{10!}$$

Again, for you calculator fanatics out there, this comes out to be 0.0000382 to 3 s.f. One thing to notice about all of these answers is that they're *really small*. That's the interesting thing about the Poisson distribution: it shows us that when events happen randomly in time, large deviations from the average rate of occurrence are actually very rare. Now for some problems to try out the new equation with:

EXERCISES

14.2.1. What are the conditions under which we can use the Poisson distribution?

14.2.2. How would the statement, "X follows Poisson distribution, with parameter 7," be written symbolically?

14.2.3. The residents of a house know that, on average, they receive five leaflets throught their letterbox per day. What is the probability that today they receive only two leaflets?

14.2.4. On average, a bakery serves 45 customers per hour. What is the probability that in any given hour they serve 60 customers?

14.2.5. A greengrocer knows that, on average, they sell 12 boxes of strawberries *every 20 minutes*. What is the probability that they will sell 40 boxes of strawberries in *an hour*?

14.2.6. A machine produces fibre optic cable. The manufacturer knows that the machine makes, on average, two dents in the cable per 10 metres. The manufacturer selects and cuts out 1 m of the cable. This 1 m can only be sold if it has no dents in it. What is the probability that the selected 1 m is sellable?

14.2.7. In a primary school, there are 30 children in a class. The teacher knows that, on average, 10% of the class will be absent from school on a given day. What is the probability that there are 29 children in school today?

14.2.8. A mobile phone company knows that, on average, 1% of the calls to its help centre will be customers wishing to terminate their contract. It receives 1000 calls on a normal day. What is the probability that 15 of its customers will call to terminate their contracts tomorrow?

14.3 Using Binomial and Poisson Models

We've already seen some reasonably straightforward applications of the binomial and Poisson distributions, but we can actually use them *together* to solve problems which involve large numbers. Firstly, we're going to look at some pieces of information that we can easily find when we know that our data is distributed in a certain way.

Mean and Variance

As we're sure that the definitions of these two things have been drilled into you, we'll keep this brief. The *mean* of a set of data is found by summing all of the pieces of data and then dividing this by the number of pieces of data in the set. The *variance* of a set of data is a measure of how much pieces of data vary (on average) from the mean. There – not too painful, was it? When we know that data is distributed in either the binomial or Poisson distribution, it's actually very, very easy to find the mean and the variance of our data. Here's how it's done:

- **Binomial**

  · *Mean:* The mean number of successes is simply the number of trials multiplied by the probability of success. In algebraic terms, $\mu = np$.

  · *Variance:* The variance of the number of successes is just the number of trials multiplied by the probability of success, then multiplied by the probability of failure. In algebraic terms, $\sigma = npq$.

- **Poisson**

  · *Mean:* We define the Poisson parameter in exactly the same way that we define the mean of the set of data, so $\mu = \lambda$.

  · *Variance:* The variance is also defined in the same way that we define the Poisson parameter of the set of data, so $\sigma = \lambda$.

Convenient, eh? We're now going to look at a handy trick that we can use when our normal approaches to problems break down.

The way that we worked with the binomial distribution before was simply to plug values into the equation:

$$P(X = r) = \binom{n}{r} p^r q^{n-r}$$

This works perfectly well for all values of n, p and q, but there are occasions where we can run into difficulties. Imagine for a moment that we were asked to solve the following problem: A lavish Las Vegas casino has introduced a new game. The probability of winning is 0.000005 and in the first year of the game being offered they expect the game to be played 1000000 times altogether. What is the probability that 30 people out of the 1000000 participants win?

If we plough straight into the problem without thinking, we're going to run into difficulties. You see, plugging these numbers into the binomial distribution equation will yield this beast:

$$P(X = 30) = \binom{1000000}{30}(0.000005)^{30}(0.999995)^{999970}$$

If you go to your calculator and ask it to do $\binom{1000000}{30}$, it will politely but firmly decline. So how can we get around this problem?

Well, we can pull a clever trick. We've just found out that when working with the binomial distribution the mean is np, and when working with the Poisson distribution the mean is simply λ. So how about using the Poisson distribution to *approximate* the solution to this problem?

As it turns out, this trick is incredibly helpful. The value of np here is $1000000 \times 0.000005 = 5$, so in our Poisson approximation, $\lambda = 5$. Now all we need to do is plug this into the Poisson formula to get:

$$P(X = 30) = \frac{e^{-5} \cdot 5^{30}}{30!}$$

Much nicer, I'm sure you (and your calculator) will agree. The only minor difficulty that we hit is *just how good is this approximation?*

As it turns out, it is stunningly good in this case. As a rule of thumb, most textbooks say this approximation is good enough when "n is large and p is small". That's nice and vague, but with any value of n greater than about 30 and p less than about 0.1, the answer you'll get from this approximation is pretty much the same as what you'd get if you slogged through the binomial method. So our $n = 1000000$ and $p = 0.000005$ are going to give an answer that is pretty much spot on – to any sensible number of significant figures. Here's one last example of using this approximation, and then on to the final set of exercises:

> There is a new, very rare disease suspected on a small island. The population of the island is 10000, and the probability that a particular person has the disease is 0.0001. The disease is not contagious, so the probability that a certain person has the disease is independent of how many other people have the disease. The government of the island only has enough medication to cure two people of the disease. What is the probability that they will need to buy more?

With numbers like these, it's clear that we're going to need to be approximating at some point, but let's take a moment to look at the question carefully. Rather than finding the probability that they *do* need to buy more medication (which would mean finding the probability that $3, 4, 5, 6, 7, 8, 9, \ldots, 9999$ or 10000 people are infected), it's much easier to find 1 minus the probability that they *don't*. For them *not* to need to buy more, either 0, 1 or 2 people can be infected.

We first need to look at the probability that no one has the disease: We can do this in the binomial way because the "choose" coefficient is going to be $\binom{10000}{0}$, which is 1. Hence the probability that no one has the disease is just

$$\binom{10000}{0} (0.9999)^{10000} (0.0001)^0 = 0.9999^{10000}$$

In looking at the probability that one or two people have it, we're going to approximate. In the binomial setup of the question, $n = 10000$ and $p = 0.0001$, so $np = 1$. Hence in our Poisson approximation $\lambda = 1$. Now using Poisson, the probability that one person has the disease is then:

$$P(X = 1) = \frac{e^{-1} \cdot 1^1}{1!} = e^{-1}$$

The probability that two people have the disease is:

$$P(X = 2) = \frac{e^{-1} \cdot 1^2}{2!} = \frac{e^{-1}}{2}$$

So, all in all, the probability that the government *will* need to buy more medication (what the initial question asked) is just:

$$1 - \left(0.9999^{10000} + e^{-1} + \frac{e^{-1}}{2}\right) = 1 - \left(0.9999^{10000} + \frac{3e^{-1}}{2}\right)$$

Like before, leaving the answer in this form is fine. Here's a final set of exercises to ensure you for having a high probability of success in "Test Youself"!

EXERCISES

14.3.1. If $X \sim B(3, 0.2)$, what is the mean?

14.3.2. If $X \sim B(5, 0.4)$, what is the variance?

14.3.3. If $X \sim Po(5)$, what is the mean?

14.3.4. If $X \sim Po(12)$, what is the variance?

14.3.5. Without giving numerical values, what are the conditions on n and p that allow us to approximate a binomial distribution to a Poisson distribution?

14.3.6. What is the main reason that we would *want* to approximate a binomial distribution to a Poisson distribution?

14.3.7. A manufacturer makes, on average, 50000 microchips per year. Thanks to their excellent quality control, the probability that any given microchip is sold faulty is 0.00008. What is the probability that there will be exactly six microchips sold faulty this year?

14.3.8. A dentist sees 1000 patients per year. The probability of any given patient requiring a new (and expensive) treatment is 0.002, and the dentist only has one course of the new treatment in stock. What is the probability that the dentist will need to buy more of it before the year is out?

Where Now?

Here, we've only looked at two of the most basic distributions – but there are plenty more. The Normal distribution *is* sometimes taught before degree level, but needs much more rigorous explanation than it is given, so let's leave that one to your lecturers. *Probability* (J. Pitman, Springer-Verlag, 1993) covers both the binomial and Poisson distributions in much greater depth than we've had chance to here, and delves into a large number of other distributions, too.

The number of problems that can be solved by modelling a situation as one of the known distributions is far wider than the selection that we have covered here. As you learn the more rigorous reasoning behind some of the topics discussed above, you'll discover the motivation for a whole new style of problem. For example, we can use a trick known as "thinning" (or "colouring") the Poisson distribution to extend what we've already covered, and this lets us do a detailed analysis of a problem where there are multiple *different* events happening through the same period of time. From that point, there are plenty of new tricks to be employed in order to solve more and more complicated problems!

15
Making Decisions

Test Yourself

If you think you are already comfortable with this material, try these questions first and mark them using the answers at the back of the book. If you get them all right, you're probably ready to move straight on to the next chapter. If some look tricky, study the chapter first and then come back to these when you're ready.

1. Describe briefly the "behavioural" (also known as "subjective") approach to probability and state why, in the real world, it is often more helpful than a "frequency" approach.

2. Describe the spinner which would be viewed as the equivalent bet to flipping two coins simultaneously, and considering two heads a success and everything else a failure.

3. What percentage of a spinner should be coloured for "success" in order for it to be equivalent to the bet "choosing a blue ball from a bag containing seven blue and three black balls."

4. What angle, in degrees, should the "success" colour of a spinner take up in order for it to be equivalent to the bet "roll a 3 on a fair six-sided die."

5. A person states that they will only play a game if, after playing many, many times, their expected loss is zero and their expected gain is zero. Describe the utility function that this person has.

E. Hurst and M. Gould, *Bridging the Gap to University Mathematics*,
DOI: 10.1007/978-1-84800-290-6_15,
© Springer-Verlag London Limited 2009

6. Use expected monetary value (EMV) strategy to decide what price should be charged for a fair game with a $\frac{1}{10}$ chance of winning and a £20 prize. (The original stake is not returned upon winning.)

7. A game costs £1 to play and offers a $\frac{1}{4}$ chance of winning a small prize and a $\frac{1}{8}$ chance of winning a large prize. It *is* possible to win both a large and a small prize at the same time, and the events "winning a small prize" and "winning a large prize" are independent. A large prize is £5, and if a prize is awarded then the original stake is not returned. How much should the small prize be so that the EMV strategy of the game results in no gain and no loss?

8. What should the probability of winning a prize be if a game is fair, costs £5 to play and rewards £42 for a win? (The original stake is not returned upon winning.)

9. A gardening centre needs to buy plants for an upcoming show. They can either buy 500 plants for £250 or 200 plants for £150. At the show, if it is sunny they expect to sell 480 plants. If it is not sunny they expect to sell 100 plants. At the end of the show all unsold plants will have to be thrown away. Plants sell for £2 each. The weather forecast predicts a 10% chance of rain for the day. Should they opt for 500 or 200 plants?

10. An estate agent needs to recruit new staff for the coming summer. If business is good in the summer, they will need ten new employees. If it is bad, they will only need two.

    They can go to recruitment agency A, who will find them up to 12 employees in time for the summer. The cost of this service is £10000. Alternatively, they can go to recruitment agency B, who will find them up to three employees in time for the summer. The cost of this service is £4000.

    Once the summer arrives, every member of staff that the company is short on they will need to recruit themselves. The cost of doing this is £2000 per employee. Any surplus employees found by agencies can be declined at no cost.

    Advise the estate agent which agency to choose.

15.1 A Whole New Probability

For those of you in a hurry, this chapter is probably the least essential in the book. Before degree level, "decision mathematics" is an optional application, and in your degree we'd be very surprised if it were compulsory in the first year. Despite this, decision analysis is an exceptionally "employable" skill in both senses of the word: we can use it often in everyday life, and it is highly sought after by those in the professional world. We'd thoroughly recommend that you see what this chapter has to offer and, if you enjoy the material, to seek out a course on decision analysis in your studies.

How Many Is Many?

"Decision mathematics" before university focuses heavily on the implementation of algorithms (is that *really* mathematics? I digress...). The decision mathematics beyond this – and indeed in the "real world" of financial markets and risk analysis – is far more concerned with making decisions in the face of some uncertainty. For this, it's helpful to take a whole new slant on the idea of probability.

If you've done some previous study of statistics, you'll know the great difficulty in coming up with a definition of "probability". We're going to look now at some different attempts to quantify this elusive character.

The *frequency* approach to probability is one that you're probably familiar with. The frequency approach is the idea that the "probability" of an event happening is equal to the number of times it happened, divided by the number of trials. Although most people are happy to accept this idea because it "seems right," statisticians have a big problem with it.

First of all, in the form just stated, it is *just plain wrong*. If I roll a die once and do not roll a 6, does that mean that the probability of rolling a 6 is $\frac{0}{1} = 0$? No, of course not, but to get around this problem we have to add the clause that makes mathematicians vomit: "When we repeat the experiment many times."

If you're not sure as to why people are so averse to this, it's the word "many" that does it. How many is many? 100? No. If we repeat the trial 100 times it's highly unlikely that exactly $\frac{1}{6}$ of the rolls will be a 6 (actually, it's impossible because 6 doesn't divide 100 without remainder, but our worries are far greater). Because of the random outcomes of the trials, we might get zero 6s, or we might even get 100 6s. So how about 1000 trials? Or 1000000? Can we be sure that after any finite number of trials our "probability" will be exactly $\frac{1}{6}$? No.

What we have to settle for is this statement, which is commonly referred to as "the golden theorem":

> When the number of trials tends to infinity, the number of successes divided by the total number of trials will tend towards the probability

of a single success, with probability 1.

Not quite so nice, eh? We actually have to use the word "probability" in our definition of the word "probability," which is hardly ideal. Besides, this "tends to infinity" business is all very theoretical, and the process of decision making is very real. Because it is impossible for us to have "infinite trials" of an event, suddenly the frequency approach to probability isn't looking so hot.

Let's take a look at a different approach: geometric. This is the idea that, for *any* event, we could make a circular spinner with different colours on it, so that the "probability" of something happening is equal to the "probability" of the spinner landing on a certain colour. So if we were making a spinner about the event "rolling a 6 on a fair, six-sided die" and we already knew that the "probability" of this (whatever that may mean) was $\frac{1}{6}$, we would make a spinner that was $\frac{5}{6}$ red and $\frac{1}{6}$ green, with green corresponding to a success. Notice that the geometric approach to probability doesn't actually make any attempt to *define* the word "probability." It is simply an *approach*.

Now that we have this idea in place we can start to look at what is really helpful to us: the *behavioural* approach to probability. This fascinating concept deals with one of the three vices of modern mankind – gambling (the other two being lust and intoxication, and being a student you'll naturally partake in neither). Interestingly enough, using gambling as our motivation produces a very useful approach to probability.

Imagine the scene: A friend offers you a bet. You win some reward if you win the bet, but you lose *nothing* if you lose. We won't worry about how much you win, we'll just assume that you are happy to bet anyway because there is something to be gained but no chance of losing anything. Your friend hasn't told you what the bet is on yet, so you sit down at a table and watch as they pull out a spinner and a fair, six-sided die. The spinner is $\frac{1}{4}$ green and $\frac{3}{4}$ red. They say:

> "You can choose which of these to play. If you play the spinner, you win if the spinner lands on green. If you choose the die, you win if you roll a 4."

You choose to play the spinner, because your chance of winning seems greater. But then your friend says:

> "Actually, I've changed my mind. I don't like that spinner. We'll use this one instead. *Now* would you rather play the spinner or roll the die?"

The new spinner that you're offered is $\frac{1}{8}$ green and $\frac{7}{8}$ red. You say you'd rather roll the die now, to which your friend says:

"Actually, I've changed my mind. I don't like that spinner. We'll use this one instead. *Now* would you rather play the spinner of roll the die?"

This process repeats about 30 times, and you notice something: If you say you'd rather roll the die, the next spinner that you're offered will always have slightly *more* green than before, and if you say you'd rather play the spinner, your friend next offers you a spinner with slightly *less* green than before. What's more, every successive time a spinner is offered, the amount of change from the previous spinner gets less and less. Eventually, the spinner that is offered to you looks to have about the same chance of landing in the green as you think you have of rolling a 4 on the die, so you say to your friend that you are indifferent between the two. At this point your friend pulls out a protractor, measures the angle of the green section of the spinner and divides this by the full 360 degrees. *This* fraction is your personal evaluation of the probability of rolling a 4 on the die, because you said that you think the probability of rolling the 4 is equal to the probability of the spinner landing on the green. In our examination of the geometric approach to probability we said that for every event we could make such a spinner – and here we've used the behavioural approach to do exactly that.

One final thing to note about the behavioural approach to probability is that it only finds a *particular person's opinion* on the probability. For an event where there is no known, fixed probability of success, different people may be indifferent between the two bets at very different points. For example, betting on a football match, person A might be indifferent between the two bets "land on green" or "team X wins" when the spinner is $\frac{1}{3}$ green, but person B may not say so until the spinner is $\frac{9}{10}$ green. For this reason we say that the behavioural approach to probability doesn't tell us *the* probability of an event occurring, instead it tells us an individual's *personal elicitation* of the probability of an event occurring.

EXERCISES

15.1.1. Briefly describe the frequency approach to probability.

15.1.2. What is the main difficulty that faces us if we are only prepared to consider the frequency approach?

15.1.3. State the golden theorem (i.e., the rigorous idea behind the frequency approach).

15.1.4. In the geometric approach, what fraction of the spinner should be coloured for "success" to represent the probability of a fair coin landing heads up?

15.1.5. In the geometric approach, what angle (in degrees) should the "success" part of the spinner take up to represent the probability of choosing a white ball from a bag of three white, four black and three grey balls?

15.1.6. An individual is indifferent between a spinner bet and the bet "team A wins a cricket match" when the "success" portion of the spinner has an angle of 54 degrees. What is their personal elicitation of the probability that team A wins the match?

15.2 Story Time

Until actually visiting a casino, Hannah found it difficult to understand why anyone ever played roulette. Games like Blackjack have been proven by some professionals to be "beatable" by a skilled player, but short of the handful of people worldwide who claim to be able to analyse the path of the ball to predict its resting place with some degree of accuracy, roulette is a bad bet for the player. In roulette there are 37 possible outcomes and a win for the player is paid at 36 to 1 (with the original stake not returned upon winning). Before actually standing at those tables, Hannah found it hard to see why people would play a game that offered those kinds of odds. Anxious to find out more, she flew to Vegas.

When she'd watched the game for just a few minutes, she had chips in one hand, wine in the other, and was in on the action. The excitement of the game, the thrill of watching the wheel slow to a halt, the prospect of winning a large sum of money for a small outlay: all of these factors added up to her loving every minute of it. You see, she enjoyed playing so much that she was prepared to excuse the slightly unfavourable odds to the player.

The reason Hannah could never see why people played the game initially was that she was viewing the bet solely in terms of expected monetary value (EMV) – she could see that for every £37 that she placed on the table, she should expect to lose £1. But when she saw the draw of the tables, she adopted a different approach: She was what decision analysts call "risk taking." Both EMV and this "risk taking" approach are examples of *utility functions*:

A *utility function* is a function that assigns a value to a reward.

What the definition is saying is that a player in a game might value the reward of that game in a "strange" way. EMV strategy is the most basic utility function:

it assigns a value exactly equal to the reward. So a player playing EMV strategy would only play roulette if the game offered 37-to-1 or better: They would never play a game where they expected to lose some money if they played many, many times.

But there are plenty of other utility functions out there. Consider the following two scenarios:

Firstly, you are at the seaside and there is a competition. It costs £1 to play. You choose a ball from a bag containing 1001 balls, 1000 of which are white and one of which is gold. If you choose a white ball, you lose your £1. If you choose the gold ball, you get your £1 back, and another £999, giving you £1000 in total.

Secondly, you spot another competition. It costs £1000 to play. You choose a ball from a bag containing 1003 balls, 1002 of which are gold and 1 of which is white. If you choose the white ball, you lose your £1000. If you choose a gold ball, you get your £1000 back, and another £2, giving you £1002 in total.

The question now is which of these games do you play? The EMV strategy works like this:

Playing game 1 1001 times, £1 per play = £1001 outlay. Expected return is 1000 losses and 1 win, so £1000. Total loss of £1.

Playing game 2 1003 times, £1000 per play = £1003000 outlay. Expected return is 1002 wins and 1 loss, so $(1002 \times 1002) - 1000 = £1003004$. Total gain of £4.

So by EMV strategy, where we simply look at what would happen if we played the game enough times for all of the possible outcomes to occur, the second game is better. But if you were walking along and saw these two games, would you really play the second one? Risking £1000 for only £2 in potential reward seems very bad, even though you are very, very unlikely to lose your £1000. The chance of gaining £999 in the first game for just £1 in outlay is very attractive, though – if you *had* to play one of the two games, wouldn't you opt for the first?

The explanation behind this logic again lies with utility functions: we *value* the first reward much more highly than we do the second, and so our utility function makes it seem more attractive. This sort of reasoning is exactly why Hannah found roulette so enticing: the reward for a win was quite large, and she didn't feel that her outlay was particularly high. People employ similar thinking every week when they play lotteries with massive jackpots: they know that their chance of winning is very low, but the reward for doing so is life changing so they *value* it highly enough to excuse the unfavourable odds.

Utility functions play a huge part in everyday life – they are pretty much the sole reason that businesses dealing with risk exist in the first place. Consider a home insurance firm. They'll charge a customer premiums that are slightly higher than they should be if they used EMV to work out the chance of the customer making a claim, multiplied by the cost to the company if they do so – and yet the customer is more than happy to pay this because if they weren't insured and got burgled, they might end up having nothing at all.

In the next section (where we look at how to solve decision problems) we'll look at working with some numerical values in utility functions, but for now here are some more theoretical exercises:

EXERCISES

15.2.1. What is a utility function?

15.2.2. At a school fundraising event there is a game that costs £1 to play. You spin a spinner and if the spinner lands on the green section, which covers $\frac{4}{10}$ths of the board, you win. Use EMV to work out how much the prize should be for doing so (including the money that you bet initially).

15.2.3. A game offers a reward of £10 to a winning player, with a $\frac{1}{50}$ chance of winning. What should the owner of the game charge people to play if he wants to *expect* to make a profit? Give your answer as an inequality.

15.2.4. After some research, the owners of the game in the previous question find that the players of the game have a utility function of 1.2 times the actual reward; that is, they enjoy the risk of the game so much that they value the prize at 1.2 times its real monetary value. What would be the best price for the owner of the game to charge now?

15.2.5. A game costs £2 to play, and offers a reward of £100 to a winning player. What probability of winning should the game have for the game to be "fair" for both the player and the owner?

15.2.6. The game in the previous question takes a very long time to play, and it turns out that players find it boring. This results in them having a utility function of 0.8 times the actual reward. What probability should the game have of winning for it still to be attractive to players, despite this "risk-averse" utility function? Give your answer as an inequality.

15.2.7. A computer owner knows that, in any given year, the probability they will need to make a claim on their insurance is $\frac{1}{30}$, and that if they do so the claim will be for £1000. They know that they will never need to claim more than once in a year. Assuming that the insurance company wants to expect to make some profit, what *monthly* premiums should the person expect to be charged? Give your answer as an inequality.

15.3 Decision Problems

Now that we have explored the ideas of probability and utility functions, we're going to end the chapter by looking at decision problems. Decision problems are problems where an individual has to choose their decision from a range of possibilities in the face of some uncertainty.

Decisions, Decisions...

Decision problems include a whole range of things, so here are some examples:

- A company choosing the pay rise to offer their employees. Too high will cost the company too much money, but too low and employees might start looking for jobs elsewhere, meaning the company will have to spend more money on recruitment.

- An individual deciding how many advertisements they would like to place in a newspaper for their new product. Too few advertisements and the product won't sell very well, but too many and the advertisements will only be reaching the same people multiple times, and still cost the individual money to place them.

- How much food to buy for a party. Too little and people will eat it all and get hungry, too much and some of it will go to waste.

If we can deal with problems like these in a numerical way, we can get a definite answer of what is a *good* decision. Here are a couple of worked examples:

An individual wishes to set up a company for printing leaflets. To do so, they will need to buy a printer. There are two such printers on the market:

- **A small printer** – this costs £2000, and can print 1000 leaflets per day.
- **A large printer** – this costs £8000, and can print 5000 leaflets per day.

Both of the printers have a lifetime of 100 days – that is, after 100 days of operation they will become useless and have no resale value. The individual is trying to get one of two different printing contracts. If they get contract A, they will be able to sell 4500 leaflets per day. If they don't get contract A, they will definitely get contract B, but will only be able to sell 800 leaflets per day. They will earn 10p per leaflet sold. Use EMV strategy to advise on the decision of which printer to buy.

Hopefully you can see the difficulty here: The individual has to choose which printer they want to buy *before* they find out which contract they're going to get. We have lots of monetary values given in the question, but we don't know the probability of getting contract A or contract B. This means that the answer we give is going to be of the form, "If you think that the probability of you getting contract A is greater than p, then buy the large printer. If not, buy the small one."

To solve a problem like this, we start by drawing a table. There are two possibilities for the decision, and also two possible outcomes in terms of which contract is awarded, and so we have four possibilities altogether:

- Buy small printer, get contract A.

- Buy small printer, get contract B.

- Buy large printer, get contract A.

- Buy large printer, get contract B.

In our table, we can put the profits earned by the individual in each of the cases. For example, in the "small printer, contract A" cell, we calculate 1000 leaflets per day, times £0.10 per leaflet, times 100 days, minus £2000 cost for the printer = £8000 profit overall. Doing a similar calculation for every cell, we get:

	Small	Large
A	8000	37000
B	6000	0

From this point, we need to go on to find the *expected* value of the profits. The only problem here is that we don't know the probability of the individual getting contract A, so let's call that x_1, and we'll call the probability that the individual gets contract B x_2.

To get our final answer we're going to need to draw a graph of two lines to do with the probabilities and see where they intersect. We know that the individual will get one of the contracts, so we know for sure that $x_1 + x_2 = 1$.

We can use this fact to do a clever trick: if we imagine that, for a given decision, $x_1 = 1$, then we know for sure that $x_2 = 0$. This is one point on one of the lines. Then if we imagine that, for the same decision, $x_1 = 0$, then we know for sure that $x_2 = 1$. This is another point on the same line. Because the line is going to be straight, now that we have these two points we can draw the line. Then we just repeat the process for the other line and we're done. That might seem a little tricky in words, so here's the working for our example:

- Decision 1: Profit $= 8000x_1 + 6000x_2$

- Decision 2: Profit $= 37000x_1 + 0x_2$

We are going to plot "profit" as the y-axis and "x_1" as the x-axis, so for the line of decision 1, we have a point at $(0, 6000)$ and another at $(1, 8000)$. For the line of decision 2, we have a point at $(0, 0)$ and another at $(1, 37000)$ (remember, when $x_1 = 0$, then $x_2 = 1$). Figure 15.1 shows the graph.

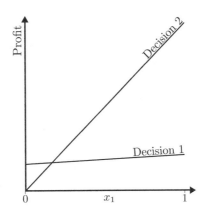

Figure 15.1

We need to find the point where the two lines intersect. To do this we're going to need the equations of the lines. For the small printer, the line starts at 6000 and increases by 2000 over the course of the graph. For the large printer, the line starts at 0 and increases by 37000 over the course of the graph. This means that our two equations are just:

$$y_1 = 6000 + 2000x_1$$
$$y_2 = 0 + 37000x_1$$

Solving simultaneously:

$$y_1 = y_2 \text{ when } 6000 + 2000x_1 = 37000x_1$$

$$35000x_1 = 6000$$

$$x_1 = \frac{6}{35}$$

So there we have it. If the individual thinks the probability of getting contract A is greater than $\frac{6}{35}$, then he or she should buy the large printer. If he or she doesn't, then the small printer should be the choice.

Have a quick review of that last example question. It's a weird combination of logic and mathematics, so it might seem a little strange. When you've mastered the basic idea of these things, they're all pretty similar, so you shouldn't have any major difficulties tackling them once you're at university and doing problems like this for assessment! Anyway, if you're happy with the previous example, here's one more, and then some exercises to round off the topic:

> A comic book seller is looking to extend their range. They have the option of buying a box of either *Superdude* or *Awesomeguy* comics, but not both. A box of 100 *Superdude* comics costs £1000, while a box of 100 *Awesomeguy* comics costs £500. At an upcoming comic fair, the seller will be assigned one of two pitches. If they are assigned the blue pitch, they expect to sell 95 of these new comics, and if they are assigned the yellow pitch they expect to sell 30 of these new comics. A *Superdude* comic will sell for £14 and an *Awesomeguy* comic will sell for £6, but at the end of the fair the seller will have to throw away any unsold comics because of lack of storage space.
>
> Just before making the purchase, the seller finds out that the probability of getting the blue pitch is $\frac{4}{10}$, and the probability of getting the yellow pitch is $\frac{6}{10}$. Which comics should he buy?

Let's start by drawing the table of profits:

	Superdude	Awesomeguy
Blue pitch	330	70
Yellow pitch	−580	−320

This means that we find the points $(0, -580)$ and $(1, 330)$ on the line of the decision to buy *Superdude,* and the points $(0, -320)$ and $(1, 70)$ on the line of the decision to buy *Awesomeguy*. These yield the equations $y_1 = -580 + 910x_1$ and $y_2 = -320 + 390x_1$. Solving simultaneously:

$$y_1 = y_2 \text{ when } -580 + 910x_1 = -320 + 390x_1$$

$$520x_1 = 260$$

$$x_1 = \frac{1}{2}$$

So he should choose to buy *Superdude* only when the probability of getting the blue pitch is greater than $\frac{1}{2}$, which it isn't here. This means that he should choose to buy *Awesomeguy*.

EXERCISES

15.3.1. A car salesman can buy one of two types of mats for their cars. A set of standard mats costs £5 and sells for £20 and a set of deluxe mats costs £8 and sells for £35. The salesman can only buy mats in boxes of 1000 sets.

If the year is a "good" year he will sell 1000 sets and if the year is a "bad" year he will sell 400 sets.

Advise the salesman which mats to buy.

15.3.2. A bank is trying to attract new customers by offering them a free gift for opening an account. They can buy either 500 such gifts for £5000, or 1000 such gifts for £9000.

If the bank attracts corporate business they expect to attract 950 new customers, but if they don't then they expect to attract only 200 new customers. They predict that the chance of attracting corporate business is $\frac{8}{10}$. If they attract a customer but cannot offer them the free gift, they will lose the customer. The bank earns £25 for every new account that is opened.

Should the bank buy 1000 of the free gifts, or just 500?

15.3.3. A new company wishes to open a stationery shop on a high street. They can either choose a large unit with rent of £1000 per week or a small unit with rent of £600 per week. If the stationery shop is popular they will make £5000 profit on their sales in the large unit, or £2000 profit on sales in the small unit (these prices exclude the rent payment). If the stationery shop is unpopular they will make only £700 profit on sales, regardless of which unit they are in.

Advise the company which unit to invest in.

Where Now?

Decision analysis moves very directly into an area of mathematics that is currently expanding rapidly: game theory. In mathematics, a game is simply any situation where two or more people make a decision simultaneously, without knowing which decision the other "player(s)" will make, and where the decisions of everyone interact. In this way, "games" are everything from poker to operations in control rooms during a war. Finding the optimal strategy to play on an opponent's weaknesses, or even finding the most cooperative strategy, are all part of game theory. Because of the complexity of many of the strategies, only the most basic "games" are explored in the first year of a course on such a topic, yet the material is still fascinating.

Outside of this, decision analysis as a skill in its own right can be extended greatly beyond what we've looked at here to include situations where there are many more than two possible choices to be made, and also to situations where more than one decision has to be made at a time. In the financial world, decision analysis is a *very* sought after skill indeed. *Game Theory: A Very Short Introduction* (K. Binmore, Oxford University Press, 2007) is an interesting text which explores the wide world of game theory and decision making.

16
Geometry

Test Yourself

If you think you are already comfortable with this material, try these questions
first and mark them using the answers at the back of the book. If you get them
all right, you're probably ready to move straight on to the next chapter. If some
look tricky, study the chapter first and then come back to these when you're
ready.

1. Is reflection in the line $y = x$ an isometry in the plane?

2. Is translation by $\begin{pmatrix} 3 \\ 1 \end{pmatrix}$ an isometry in the plane?

3. Are the triangles in Figure 16.1 congruent?

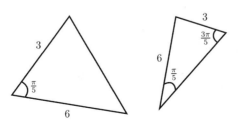

Figure 16.1

E. Hurst and M. Gould, *Bridging the Gap to University Mathematics*,
DOI: 10.1007/978-1-84800-290-6_16,
© Springer-Verlag London Limited 2009

4. Are the triangles in Figure 16.2 congruent?

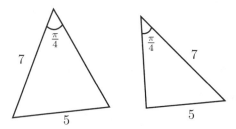

Figure 16.2

5. Find angle x in Figure 16.3, given that O is the centre of the circle.

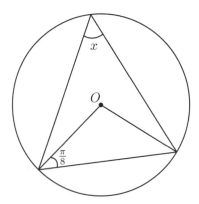

Figure 16.3

6. *Prove* that for any circle, the angle from an arc to the centre is twice the angle from that arc to the edge.

7. Find x in Figure 16.4.

8. *Prove* that for any triangle:
$$\frac{a}{\sin A} = \frac{b}{\sin B} = \frac{c}{\sin C} = 2R$$
where R is the radius of the circle whose circumference meets all three corners of the triangle.

9. What is the sum of the angles of an equilateral triangle on the sphere, one of whose sides is part of an equator and one of whose points lies at the "north pole"?

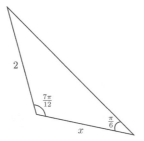

Figure 16.4

10. What range of values could we get when summing the angles of a triangle drawn on an upper hemisphere?

16.1 Old Problems, New Tricks

One thing that is sadly absent from pre-university mathematics is Euclidean geometry. In both pure and applied mathematics, geometry lies at the very heart of many problems, yet most people come to university knowing no more than how to "spot" when we can use certain rules on angles in a circle and roughly how to work with triangles. In this chapter we're going to *prove* some of the things that you've been happily applying for some years, and also look at some 3-dimensional geometry.

Transformations and Isometries

As we're sure you're aware, there are four basic transformations that we can apply in two dimensions (matrix representations of these were explored in the chapter on matrices as maps). They are:

- **Translation** – "moving" things, but keeping them at the same orientation and size.

- **Dilation** – also known as "enlargement," this is about making things bigger or smaller.

- **Reflection** – taking the "mirror image," through some given line of reflection.

- **Rotation** – "spinning" a shape by a certain amount, around a given centre of rotation.

With these tools, we can move and change shapes in all kinds of ways. For this reason, we have the word *isometry* to describe any transformation that preserves the distance between any two points. This means that translation, rotation and reflection are all isometries in the plane, but enlargement is not. Lots of the geometry studied in a degree course relies heavily on this idea, so learning this definition now is definitely time well spent:

The map f of the plane is an *isometry* if for every two points A, B:

$$\text{distance}(A, B) = \text{distance}(f(A), f(B))$$

Just have a quick think about what the definition is saying. There's nothing too spectacular going on here.

Congruence

Before university, it is good enough to know that two triangles are congruent if you can move, flip and rotate one of them and get it exactly on top of the other one. This is definitely a good way to visualise what's going on, so don't lose sight of it, but there is a sneaky "cheat" way to check if two triangles are congruent.

If two triangles meet any of the following three congruence criteria, then they are congruent:
- **SAS** (side-angle-side)
- **ASA** (angle-side-angle)
- **SSS** (side-side-side)

Now, these will probably look a bit mystifying, so here's a guide to what they mean: If you have two triangles, and tracing in a *single direction*, you see that both triangles are identical in one of the above ways, then the triangles are congruent. Don't worry if that doesn't seem totally clear – here are some examples:

Are the triangles in Figure 16.5 congruent?

Here we have an example of the SAS congruence criteria: side-angle-side. If we read clockwise around triangle ABC, starting with side AC, we get $7, \frac{\pi}{3}, 6$ as our SAS. If we read anti-clockwise around triangle DEF, starting with side DF, we get $7, \frac{\pi}{3}, 6$ as our SAS. These are the same, so our triangles are definitely congruent. Notice that it didn't matter that we went around each triangle in a different direction: all that was important was that we went in *some* direction and didn't "jump around." Here's one last example to illustrate this:

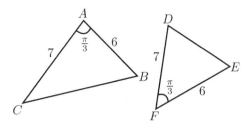

Figure 16.5

Are the triangles in Figure 16.6 congruent?

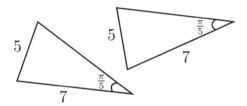

Figure 16.6

Trace around either triangle and you get $5, 7, \frac{\pi}{5}$ from a clockwise trace or $\frac{\pi}{5}, 7, 5$ from an anticlockwise trace. This is SSA or ASS respectively. Note that neither of these were criteria for congruence, so we *can't* say that our triangles are congruent. We can't say that they definitely aren't, because they still might be, but the important thing is that we can't say that they definitely *are*.

EXERCISES

16.1.1. Is rotation by $\frac{\pi}{4}$ about the origin an isometry?

16.1.2. Is dilation, scale factor 2, about the origin, an isometry?

16.1.3. Is rotation by $\frac{\pi}{4}$ about the origin *then* dilation, scale factor 2, about the origin, an isometry?

16.1.4. I reflect a shape in the line $y = x$, then dilate it by scale factor z, about (z, z). This composite transformation *is* an isometry. Find z.

16.1.5. Are the pairs of triangles in Figure 16.7 congruent, not congruent or can we not determine for certain?

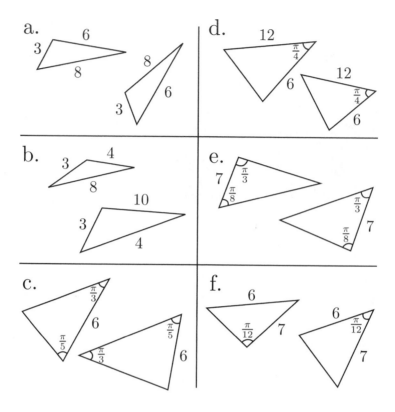

Figure 16.7

16.2 Proof

As promised in the opening of the chapter, we're now going to look at some formal proof.

The Sine Rule

At this point in your mathematical career, it's fairly certain that you're familiar with this bad-boy:

$$\frac{a}{\sin A} = \frac{b}{\sin B} = \frac{c}{\sin C}$$

We would also bet that you don't know where on Earth it actually *comes from* – or also that, in proving it, we actually get a "bonus bit" for free, making the rule:

$$\frac{a}{\sin A} = \frac{b}{\sin B} = \frac{c}{\sin C} = 2R$$

Asserting that the sine rule "must" be true simply because it happens to work on all of the triangles we've tested it on is not good enough for us fussy mathematicians. What we want is a *formal proof* of what's going on, without initially assuming the rule to be true.

Proof invariably plays a major role in university exams, but you'll quickly come to learn that *memorising* proof is no good. University exam papers have an annoying tendency to be written by some rather clever people, and so they'll have you adapt a proof that you already know to fit a new situation, not just ask you to write a proof out. For this reason it's crucial that you *understand* proof and not just remember it.

An Old Friend

Before we go headfirst into the sine rule, we're going to prove something that we'll need later. This way, we can be sure that when we use it, we're not opening a window for error to creep in. Here it is:

> The angle subtended from a chord to the centre of a circle is twice the
> angle subtended from that chord to anywhere on the edge of the circle,
> on the same side as the centre.

Again, this is no doubt something that you've been using for years without knowing a formal proof, but make sure that you're careful of that final clause: "on the same side as the centre." If we try to work with something like Figure 16.8, where we're joining our chord to the edge of the circle on the *opposite* side of the chord to the centre, we run into problems (we can see by eye that the angle at the centre is certainly not twice the angle at the edge).

So, being careful to construct the diagram correctly, in our proof we must be in one of the two situations shown in Figure 16.9.

We're going to progress by taking the diagram on the left, but if you head to "Where Now?" you'll see that a short extension to the proof ensures that it's valid for the diagram on the right too. Take a look at Figure 16.10 and then let's go!

Draw a dotted line from where the two lines from the chord meet the edge of the circle, through the centre of the circle, to the opposite side. Label the diagram as in Figure 16.11, and note that we can be sure that the angles labelled y are both equal because the triangle that they are in is isosceles.

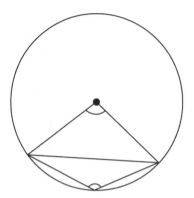

Figure 16.8

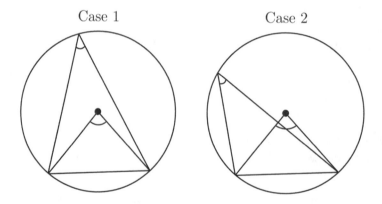

Figure 16.9

Now the proof:

1. $x + 2y = \pi$ (angles in a triangle add up to π).

2. $x + z = \pi$ (angles on a straight line add up to π).

3. So $2y = z$.

Figure 16.12 is the same diagram again, but this time with the other half labelled.

So by the same logic:

1. $a + 2b = \pi$.

2. $a + c = \pi$.

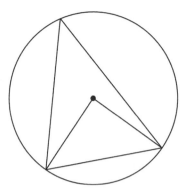

Figure 16.10

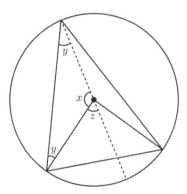

Figure 16.11

3. So $2b = c$.

Combining these:

1. $2y = z$ and $2b = c$.

2. So $z + c = 2y + 2b$.

3. Hence $z + c = 2(y + b)$

The angle at the centre is $(z + c)$. The angle at the edge is $(y + b)$. There we have it!

Now we've done that, let's take a look at a step-by-step proof of the sine rule. You'll probably need this proof in your first course on geometry, so now is as good a time as any to really get to grips with it.

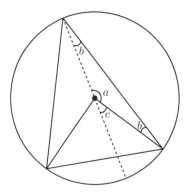

Figure 16.12

Start by drawing a circle, and draw a chord on the circle, like in Figure 16.13.

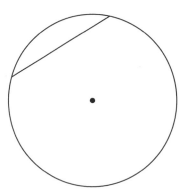

Figure 16.13

Now join each end of the chord to the centre, and draw the line from the centre to the chord that meets at right angles, as shown in Figure 16.14.

These two triangles are congruent: We had an isosceles triangle and sliced it down its line of symmetry, so now we have the SAS (or the ASA) congruence criterion. Consider the whole length of the chord to be a, so each of the smaller triangles have base $\frac{a}{2}$. By the triangle-labelling convention of using the same letter for a side and the opposite angle, the two points where the chord meets the circle are B and C, as shown in Figure 16.15.

Mark a point anywhere on the "longer" part of the circle, and join this point to B and C (because we used "Case 1" in the previous proof, we're choosing

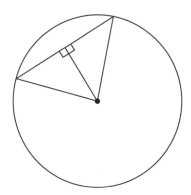

Figure 16.14

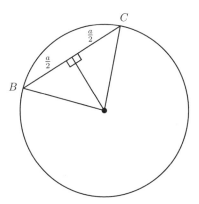

Figure 16.15

our point so that the diagram we work with looks like that. Choosing a point that would instead place us in "Case 2" of the above proof is fine too; remember we explore this case in "Where Now?"). Now we can also finish labelling the triangle in Figure 16.16, including the points D and O that we've added for reference.

Here's the logic. Make sure that you follow every step!

1. BOC is isosceles, so $2B\hat{O}D = B\hat{O}C$.

2. BOD is a right-angled triangle, so we can use the fact that BO is the hypotenuse to deduce $BD = BO \sin B\hat{O}D$.

3. BO is the radius of the circle, so we can write $BD = R \sin B\hat{O}D$.

4. $BD = \frac{a}{2}$, so we can write $\frac{a}{2} = R \sin B\hat{O}D$.

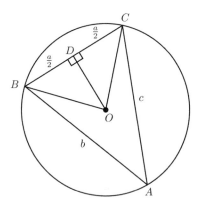

Figure 16.16

5. $a = 2R \sin B\hat{O}D$.

6. From the proof we've just done, the angle at the centre is twice the angle at the edge, so $B\hat{O}C = 2B\hat{A}C$.

7. $B\hat{O}C = 2B\hat{O}D$ (same logic as step 1).

8. $2B\hat{A}C = 2B\hat{O}D$, so $B\hat{A}C = B\hat{O}D$.

9. From step 5, $a = 2R \sin B\hat{A}C$.

10. $a = 2R \sin A$, so $\frac{a}{\sin A} = 2R$.

Finally, imagine instead that we had chosen AC as our line to bisect. Then we would get $\frac{b}{\sin B} = 2R$. If we had chosen AB, we would have gotten $\frac{c}{\sin C} = 2R$. And there we have it: $\frac{a}{\sin A} = \frac{b}{\sin B} = \frac{c}{\sin C} = 2R$. If you're anxious to try it out again fully on your own, that's great; it works every time. Just in case you're still wondering what the bonus R actually stands for, it's the radius of the circle that we draw around the triangle. The circle isn't just there to look pretty after all!

EXERCISES

16.2.1. Find angle x in Figure 16.17.

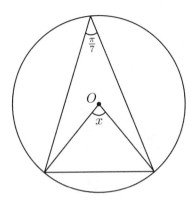

Figure 16.17

16.2.2. Which pair of angles are equal in Figure 16.18?

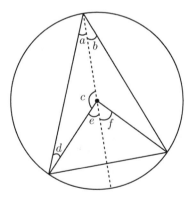

Figure 16.18

16.2.3. Which angle is $a + b$ equal to in Figure 16.19?

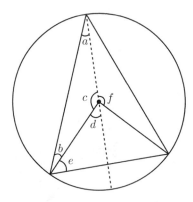

Figure 16.19

16.2.4. What is equal to $OC \sin C\hat{O}D$ in Figure 16.20?

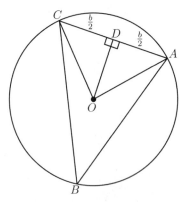

Figure 16.20

16.2.5. Write the necessary facts to show:

$$\frac{b}{\sin B} = 2R$$

16.2.6. Draw a new diagram and show that $\frac{c}{\sin C} = 2R$.

16.3 3D Geometry

All of the geometry that you've dealt with so far will probably have been 2-dimensional geometry "on the plane." But what happens if we take some of the things that we know "work" in the plane and then try them on some non-flat spaces, like the sphere? Surely everything works the same, right? *Wrong.*

Challenge Everything

Let's look at a fact that we all take for granted: "The angles in a triangle always add up to π." Now, imagine the Earth as a perfect sphere and, upon your imaginary Earth, draw the equator – but only $\frac{1}{4}$ of the way around. Join the left end of this line to the North Pole. Then also join the right end to the North Pole, so now the left and right ends meet. We have a triangle on a sphere.

The angle between the equator and the left line to the North Pole must be $\frac{\pi}{2}$ because if we look at our imaginary Earth with that point facing us, we would see what's depicted in Figure 16.21.

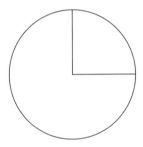

Figure 16.21

Similarly, the angle between the equator and the right line to the North Pole must be $\frac{\pi}{2}$. Now imagine looking down at the Earth from directly above the North Pole. If your imaginary picture is good enough, you're no doubt already thrilled that we see what's shown in Figure 16.22!

So the angle here is also $\frac{\pi}{2}$. Summing these angles, we get a total of $\frac{3\pi}{2}$: most definitely warning us that we're going to need to be *mighty careful* about what we *think* we know will happen on a sphere.

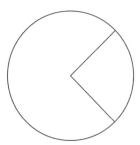

Figure 16.22

They All Laughed at Christopher Columbus...

One of the most fundamental ideas in geometry on the plane is that of the *straight line*. Given any two points in the plane, the straight line from one to the other offers us the shortest distance – and given any two points, we can *always* find such a straight line. But how about on the surface of a sphere? If we want to find the shortest distance between any two points, we're going to need a line that *curves with the sphere* because there's no such thing as a straight line on the surface of a sphere!

As it turns out, there *is* a type of line that will always offer us the shortest distance between two points on the surface of a sphere. This type of line is an arc of something called a *great circle* – but what's so great about them?

If we imagine the Earth to be a perfect sphere, then the equator is a great circle – as are any of the circles that would pass through both the North and South poles. Intuitively, a great circle is a circle which has the same radius as the sphere it is drawn upon. More formally, a great circle is the intersection of the sphere with a plane which passes through its centre, as shown in Figure 16.23.

Great circles have a wide range of uses in various geometric constructions, and at university they'll be a vital cornerstone in a whole host of proofs and problems. They're a prime example of how working with geometry on the sphere requires you to go right "back to basics," and develop a whole new intuition – and what better way to hone such intuition than to try some exercises?

EXERCISES

16.3.1. What is the sum of the angles of an equilateral triangle on the sphere?

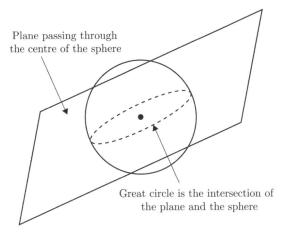

Plane passing through
the centre of the sphere

Great circle is the intersection of
the plane and the sphere

Figure 16.23

16.3.2. What is the (theoretical) minimum of the sum of the angles of a triangle drawn on a sphere?

16.3.3. What is the (theoretical) maximum of the sum of the angles of a triangle drawn on a sphere?

16.3.4. From our study of triangles, there is a way to find the sum of the angles in a quadrilateral, pentagon, hexagon, ... (any polygon) on a sphere. What is it?

16.3.5. We can only use the transformation of rotation on the sphere if we choose our axis of rotation carefully. What property must it have?

16.3.6. We can only use the transformation of reflection on the sphere if we choose our axis of reflection carefully. What property must it have?

16.3.7. The radius of the Mars is 3000 km, to one significant figure. What would the diameter of any great circle on the surface of Mars be (assuming that Mars is a perfect sphere)?

16.3.8. The volume of a perfectly spherical modern art statue is $\frac{4}{3}\pi\mathrm{m}^3$. The statue's owner wants to draw a line between the very top of the statue and the very bottom of the statue. He doesn't mind if the line twists and turns, but he wants the line to be as short as possible. How long should the line be?

Where Now?

As we've seen, geometry at university is often more about *proving* things are true than it is about getting a "toolbox" of new tricks. For this reason, lots of people find geometry courses quite difficult. Having said that, there are plenty of interesting things out there to prove: the cosine rule, Pythagoras' theorem, even the fact that the angles in a triangle always add up to π. Geometry is part of the very foundations of mathematics, with people such as Euclid working on it "back in the day," and so it's definitely something worth getting to grips with *because it turns up all over the place.*

If you think back to the proof of "the angle subtended from a chord to the centre of a circle is twice the angle subtended from that chord to anywhere on the edge of the circle, on the same side as the centre," we only explored one of two possible cases. For our proof to be complete, we need to show that it's valid in the other case too. This follows by another rule that you'll have no doubt used many times before:

> The angle subtended from a chord to the edge of a circle is the same anywhere on the edge of the circle (so long as we choose the two points of the edge on the same side of the chord).

Visually, we're saying that in Figure 16.24 angles A and B are always equal.

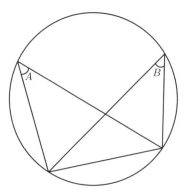

Figure 16.24

Hopefully, you'll see how we can use this fact to cover case 2 of our proof: we know that we can prove case 1, and then using this fact we see that the angle at the edge in either case is *the same*, so we get case 2 for free! If we want to be really rigorous, we need to prove, "The angle subtended from a chord to the edge of a circle is the same anywhere on the edge of the circle (so long as we choose the two points of the edge on the same side of the chord.)" We'll leave this one to you!

Modern Geometry with Applications (G. Jennings, Springer-Verlag, 1994) covers a large amount of geometry. It proves some familiar things in a rigorous way, and introduces lots of new, exciting ideas too.

If you enjoyed thinking about geometry on the sphere, try this one for size:

> Imagine two different great circles drawn on the same sphere. What can we say about where these great circles cross?

If you have a good mental picture, it becomes apparent that they'll *always* have to intersect *twice*, and the two points of intersection will also have a special property. They'll be directly opposite each other on the sphere (we call these *antipodal points* on a sphere) and it's impossible to find a pair of "parallel" great circles on a sphere: any two distinct great circles will intersect twice!

17
Hyperbolic Trigonometry

Test Yourself

If you think you are already comfortable with this material, try these questions first and mark them using the answers at the back of the book. If you get them all right, you're probably ready to move straight on to the next chapter. If some look tricky, study the chapter first and then come back to these when you're ready.

1. Find $\coth 3$, leaving your answer in terms of e.

2. Find x, given that $\cosh x = 2$.

3. Find x, given that $\tanh x = \frac{1}{3}$.

4. Find $\frac{d}{dx}(2\sinh(4x))$.

5. Find $\frac{d}{dx}(\ln \sinh(3x))$.

6. Find $\frac{d}{dx}(e^{2x}\sinh x)$.

7. Find $\frac{d}{dx}(e^{\tanh x})$.

8. Find $\int \tanh^2 x \, \operatorname{sech}^2 x \, dx$.

9. Find $\int 2\cosh(3x) \, dx$.

10. Find $\int \sinh^2(3x) \, dx$.

E. Hurst and M. Gould, *Bridging the Gap to University Mathematics*,
DOI: 10.1007/978-1-84800-290-6_17,
© Springer-Verlag London Limited 2009

17.1 Your New Best Friends

For the majority of your mathematical career to date, you'll have known the ins and outs of the trigonometric functions. Through good times and bad, these functions will have served you well over the years. There's no replacement for them, and you can't do without them. But now it's time to be adulterous.

Pronounce and Annunciate

The hyperbolic trigonometric functions have very strong similarities with the standard trigonometric functions of sine, cosine, tangent, cosecant, ... and so on. Whereas the standard trigonometric functions are based on circles, hyperbolic trigonometric functions are based on hyperbolae (graphs with an equation of the form $\frac{x^2}{a^2} - \frac{y^2}{b^2} = 1$). In order to write a standard trigonometric function in explicit terms of x we need to use an infinite sum, but hyperbolic trigonometric functions are a whole lot more friendly: these guys are simply written in terms of e^x and e^{-x}.

When we're writing about hyperbolic trigonometric functions we simply write the standard corresponding trigonometric function with an additional h on the end. For example, the hyperbolic form of $\sin x$ is $\sinh x$ and the hyperbolic form of $\cos x$ is $\cosh x$. There are also standard pronunciations of these new words, which are as follows:

Hyperbolic Function	Phonetic Pronunciation
sinh	"sinch" or "shine"
cosh	"cosh"
tanh	"tanch" or "than"
cosech	"cosech"
sech	"sech" or "sheck"
coth	"coth"

e Talking

You can relax now: the wait is over. Here are the hyperbolic functions:

$$\cosh x = \frac{e^x + e^{-x}}{2}$$

$$\sinh x = \frac{e^x - e^{-x}}{2}$$

Just like in standard trigonometry, we find $\tanh x$ by calculating $\frac{\sinh x}{\cosh x}$. This means that:

$$
\begin{aligned}
\tanh x &= \frac{\sinh x}{\cosh x} \\
&= \frac{\frac{e^x - e^{-x}}{2}}{\frac{e^x + e^{-x}}{2}} \\
&= \frac{e^x - e^{-x}}{e^x + e^{-x}}
\end{aligned}
$$

Some people prefer the expression $\tanh x = \frac{e^{2x} - 1}{e^{2x} + 1}$, which is arrived at by multiplying both numerator and denominator by e^x.

Also, just like in normal trigonometry, we see that:

$$
\operatorname{cosech} x = \frac{1}{\sinh x} = \frac{2}{e^x - e^{-x}}
$$

$$
\operatorname{sech} x = \frac{1}{\cosh x} = \frac{2}{e^x + e^{-x}}
$$

$$
\coth x = \frac{1}{\tanh x} = \frac{e^x + e^{-x}}{e^x - e^{-x}} = \frac{e^{2x} + 1}{e^{2x} - 1}
$$

Let's take a quick look at some basic questions to do with these new functions:
Find the value of $\cosh 3$.

$$
\cosh 3 = \frac{e^3 + e^{-3}}{2}
$$

Now for something a little bit more demanding:
Find the value of x if $\sinh x = 1$.

$$
\frac{e^x - e^{-x}}{2} = 1
$$
$$
e^x - e^{-x} = 2
$$
$$
e^x - e^{-x} - 2 = 0
$$
$$
e^{2x} - 2e^x - 1 = 0
$$

In case you didn't spot it, all that happened in the final step was a multiplication through by e^x. The reason that we did this is because now the equation is a quadratic equation in e^x, which is very handy indeed. Let's proceed:

$$e^x = \frac{2 \pm \sqrt{4+4}}{2}$$
$$= \frac{2 \pm 2\sqrt{2}}{2}$$
$$= 1 \pm \sqrt{2}$$

The negative root is not possible because e^x is always positive, so we discard it. Taking natural logarithms of both sides:

$$\ln e^x = \ln(1 + \sqrt{2})$$
$$x = \ln(1 + \sqrt{2})$$

Notice how we set ourselves up to be able to use the quadratic formula: this is a trick that is common in this sort of question.

Graphs

As hyperbolic functions are all expressed in terms of e^x and e^{-x}, knowing what the graphs of these functions look like should aid us in our new discoveries. Figure 17.1 shows the basic graph of e^x and e^{-x}, and we're going to use this to help work out everything else.

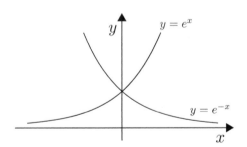

Figure 17.1

Notice that for $x < 0$, $e^x < e^{-x}$ and for $x > 0$, $e^x > e^{-x}$. This means that $e^x - e^{-x}$ will be negative for $x < 0$ and positive for $x > 0$, so when $x = 0$, $e^x - e^{-x} = 0$. Using this information, we can formulate the graph of $y = \sinh x$, as shown in Figure 17.2.

If we think about the behaviour of the graphs, then it becomes apparent that $e^x + e^{-x}$ must be symmetrical about the y-axis, and therefore $\cosh x$ will

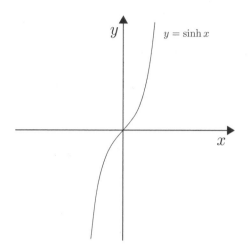

Figure 17.2

be. When $x = 0$, we see that $e^x + e^{-x} = 2$, and so $\cosh x = 1$. Using these facts, we can draw the graph of $\cosh x$, as shown in Figure 17.3.

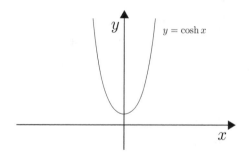

Figure 17.3

Using the fact that $\tanh x = \frac{\sinh x}{\cosh x}$ we can build up the picture of $\tanh x$, as shown in Figure 17.4. There's also a more algebraic exploration of why the curve looks like it does in the "Where Now?" section at the end of the chapter.

If you're partial to a bit of curve sketching, you may like to try the graphs of $\operatorname{cosech} x, \operatorname{sech} x$ and $\coth x$ for yourself.

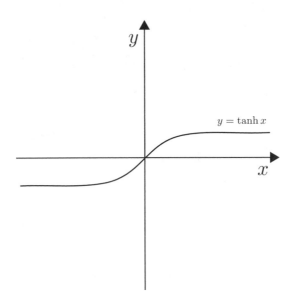

Figure 17.4

EXERCISES

17.1.1. Find the value of $\sinh 5$, giving your answer in terms of e.

17.1.2. Find the value of $\cosh \frac{3}{2}$, giving your answer in terms of e.

17.1.3. Find the value of $\tanh 2$, giving your answer in terms of e.

17.1.4. Find the value of $\coth \sqrt{2}$, giving your answer in terms of e.

17.1.5. Find the value of x, if $\sinh x = 3$.

17.1.6. Find the value of x, if $\cosh x = 4$.

17.1.7. Find the value of x, if $\tanh x = \frac{1}{2}$.

17.1.8. Find the value of x, if $\coth x = 6$.

17.2 Identities and Derivatives

Now that we're well acquainted with the hyperbolic functions we can start to look at some of the different ways in which we can manipulate them. Just like the standard trigonometric functions, there are some useful identities that we can employ in order to aid solving problems. These identities are reasonably similar to the standard trigonometric identities, so they are quite easy to remember but also quite easy to get confused!

The first identity that we're going to look at is this:

$$\cosh^2 x - \sinh^2 x = 1$$

We can find a proof of this by expressing the left-hand side of the identity as the difference of two squares, and proceeding as follows:

$$\begin{aligned}
\cosh^2 x - \sinh^2 x &= (\cosh x + \sinh x)(\cosh x - \sinh x) \\
&= \left(\frac{e^x + e^{-x}}{2} + \frac{e^x - e^{-x}}{2}\right)\left(\frac{e^x + e^{-x}}{2} - \frac{e^x - e^{-x}}{2}\right) \\
&= \left(e^x\right)\left(e^{-x}\right) \\
&= e^0 \\
&= 1
\end{aligned}$$

From this identity we can immediately find another, simply by dividing through by $\cosh^2 x$:

$$\frac{\cosh^2 x}{\cosh^2 x} - \frac{\sinh^2 x}{\cosh^2 x} = \frac{1}{\cosh^2 x}$$

Simplifying this gives us the identity:

$$1 - \tanh^2 x = \text{sech}^2 x$$

We can get another identity by taking our original $\cosh^2 x - \sinh^2 x = 1$ and instead dividing through by $\sinh^2 x$. This yields:

$$\frac{\cosh^2 x}{\sinh^2 x} - \frac{\sinh^2 x}{\sinh^2 x} = \frac{1}{\sinh^2 x}$$

Simplifying this gives:

$$\coth^2 x - 1 = \text{cosech}^2 x$$

Questions to do with hyperbolic trigonometry often involve proving a particular identity. Here's a worked example:

Show that $\cosh x \sinh y + \cosh y \sinh x = \sinh(x + y)$.

Taking the left-hand side:

$$\text{LHS} = \left(\frac{e^x + e^{-x}}{2}\right)\left(\frac{e^y - e^{-y}}{2}\right) + \left(\frac{e^y + e^{-y}}{2}\right)\left(\frac{e^x - e^{-x}}{2}\right)$$

$$= \frac{1}{4}(e^x e^y + e^{-x}e^y - e^x e^{-y} - e^{-x}e^{-y}) + \frac{1}{4}(e^x e^y + e^x e^{-y} - e^{-x}e^y - e^{-x}e^{-y})$$

$$= \frac{1}{4}(2e^x e^y - 2e^{-x}e^{-y})$$

$$= \frac{1}{2}(e^{x+y} - e^{-(x+y)})$$

$$= \sinh(x + y)$$

Just as you've probably seen in your work on standard trigonometry, if we let $x = y$ we can proceed to find another important identity from what we've just done:

$$\sinh(x + x) = \cosh x \sinh x + \cosh x \sinh x$$
$$\sinh(2x) = 2\cosh x \sinh x$$

Using similar logic to the process that we've just been through, we can derive the identity:

$$\cosh(x + y) = \cosh x \cosh y + \sinh x \sinh y$$

And from here we can also see that:

$$\cosh(2x) = \cosh^2 x + \sinh^2 x$$

Finally, we can also use the fact that $\tanh(2x) = \frac{\sinh(2x)}{\cosh(2x)}$ to derive the final identity of the section:

$$\tanh(2x) = \frac{2\tanh x}{1 + \tanh^2 x}$$

You may have noticed that the hyperbolic trigonometric identities look strikingly similar to the *standard* trigonometric identities, and they are indeed linked by something called *Osborn's rule*, which you'll find out more about at university.

Differentiating

Differentiating hyperbolic functions isn't too much of a struggle, thanks to the fact that $\frac{d}{dx}(e^x) = e^x$ and $\frac{d}{dx}(e^{-x}) = -e^{-x}$. By doing some careful differentiating, we can quickly arrive at the following:

$$\frac{d}{dx}(\sinh x) = \frac{e^x + e^{-x}}{2} = \cosh x$$

$$\frac{d}{dx}(\cosh x) = \frac{e^x - e^{-x}}{2} = \sinh x$$

These derivatives are quite common, so make sure that you know the results from memory: when we differentiate $\sinh x$ we get $\cosh x$, and vice versa. There is no minus sign when differentiating $\cosh x$ as there is with $\cos x$, so make sure that you don't get that confused!

In differentiating $\tanh x$, we're going to need the quotient rule for differentiation. Just in case you've forgotten it, this is:

$$\frac{d}{dx}\left(\frac{u}{v}\right) = \frac{v\frac{du}{dx} - u\frac{dv}{dx}}{v^2}$$

Using this on $\tanh x$ works like this:

$$\frac{d}{dx}(\tanh x) = \frac{d}{dx}\left(\frac{\sinh x}{\cosh x}\right)$$
$$= \frac{\cosh^2 x - \sinh^2 x}{\cosh^2 x}$$
$$= \frac{1}{\cosh^2 x}$$
$$= \operatorname{sech}^2 x$$

The trick in the second from last line came from the first identity that we looked at: $\cosh^2 x - \sinh^2 x = 1$. If you didn't spot that, go back and learn the identities properly!

The last thing that we're going to look at before the exercises requires the use of the chain rule for differentiation. It's not too taxing, so let's launch straight into the worked example:

Find the derivative of $\sinh(x^2)$.

Hopefully you'll be familiar with this sort of thing. The way to tackle it is to use the chain rule. The result that's required is simply:

$$\frac{d}{dx}(\sinh(x^2)) = 2x\cosh(x^2)$$

EXERCISES

17.2.1. Prove that $\cosh(x+y) = \cosh x \cosh y + \sinh x \sinh y$

17.2.2. Show that $\tanh(2x) = \frac{2\tanh x}{1+\tanh^2 x}$, using the expressions for $\sinh(2x)$ and $\cosh(2x)$.

17.2.3. Differentiate each of the following with respect to x:

    a) $\sinh(3x)$

    b) $2\operatorname{sech} x$

    c) $e^x \cosh x$

    d) $\tanh(x^2)$

    e) $x\operatorname{cosech}(\ln x)$

    f) $\ln(\cosh x)$

    g) $e^{\sinh x}$

    h) $\coth x \sin x$

    i) $e^{\cosh x}\operatorname{cosech} x$

17.3 Integration

After the marathon of identities and derivatives, we're going to wind down the chapter by looking at some integration. Looking back at what we've just done, it should be clear that:

$$\int \cosh x \ dx = \sinh x + c$$

$$\int \sinh x \ dx = \cosh x + c$$

When working with this sort of integration, remember that it's important to check that what you give as an answer would differentiate back to what you started with. Mistakes are very easy to make in this sort of situation! Here's a couple of worked examples of integration:

$$\int \sinh(3x)dx = \frac{1}{3}\cosh(3x) + c$$

$$\int \frac{1}{2} \cosh\left(\frac{x}{4}\right) dx = 2\sinh\left(\frac{x}{4}\right) + c$$

Just like when integrating standard trigonometric functions, we can sometimes make life easier by using an identity before we start with the calculus. Here's an example:

$$\text{Find } \int \sinh^2 x \, dx$$

By rearranging the identity $\cosh^2 x - \sinh^2 x = 1$, we can see that $\cosh^2 x = 1 + \sinh^2 x$. Also, we already know that $\cosh(2x) = \cosh^2 x + \sinh^2 x$. We can combine these two identities to derive the fact that $\cosh(2x) = 1 + 2\sinh^2 x$, which we can then rearrange to help us solve the problem:

$$\int \sinh^2 x \, dx = \frac{1}{2} \int (\cosh(2x) - 1) dx$$
$$= \frac{1}{2} \left(\frac{1}{2}\sinh(2x) - x\right) + c$$
$$= \frac{1}{4}\sinh(2x) - \frac{x}{2} + c$$

We use a similar trick for finding the integral of $\cosh^2 x$, using the identity $\cosh(2x) = 2\cosh^2 x - 1$.

One final thing to keep an eye out for is what we like to call "integration by eye": sometimes it is quicker to try to find a function whose derivative is the function that you're trying to integrate. Here's one final example to illustrate this:

$$\text{Find } \int \cosh^2 x \sinh x \, dx$$

Gut instinct may tell you to dive into an integration by parts, but there is a much quicker way of solving this problem. It only takes a moment to realise that $\frac{d}{dx}(\cosh^3 x) = 3\cosh^2 x \sinh x$, and what we require is simply one-third of this. That means that the answer to our problem is simply $\frac{1}{3}\cosh^3 x$: a much neater approach than getting lost in a sea of us and vs.

EXERCISES

17.3.1. Evaluate $\int \cosh(2x) \, dx$.

17.3.2. Evaluate $\int 2\sinh(4x) \, dx$.

17.3.3. Evaluate $\int 2e^x \cosh x \, dx$.

17.3.4. Evaluate $\int e^{-x}(e^{2x} + 1)\tanh x \, dx$.

17.3.5. Evaluate $\int 9\text{sech}^2(3x)\ dx$.

17.3.6. Evaluate $\int 2\cosh^2 x\ dx$.

17.3.7. Evaluate $\int \sinh^2 x \cosh x\ dx$.

17.3.8. Evaluate $\int \tanh(3x)\text{sech}(3x)\ dx$.

17.3.9. Evaluate $\int 4\tanh^3 x\ \text{sech}^2 x\ dx$.

Where Now?

As I'm sure you'll appreciate, learning *what* the sine, cosine and tangent "family" *are* is only the first step towards using them. You'll need some practice in a wide range of situations before you become fully comfortable with the hyperbolic trigonometric functions, but simply knowing of their existence and roughly how to manipulate them will be a massive help for when it comes to your first course in *properly* using them at university. *Algebra and Geometry* (A. Beardon, Cambridge University Press, 2005) is a great place to look if you're hungry for more information on the hyperbolic trigonometric functions.

From looking at conic sections to solving awkward integrals, sinh, cosh and tanh will be like old friends to you in no time. You'll be encountering them in a huge variety of situations in your degree; you'll even be needing them when you delve deeply into the integration of complex functions! There's a lot of ground to be covered but you'll definitely get a lot of mileage out of knowing these functions well.

Finally, here's a more "algebraic" exploration of the $\tanh x$ curve: As $\tanh x = \frac{e^{2x}-1}{e^{2x}+1}$, we can see that when $x = 0$ we have $e^{2x} - 1 = 0$, and so $\tanh x = 0$ at this point. To get an idea of the shape of the graph of $\tanh x$, we're going to need to look at the general behaviour. As x tends to $-\infty$, e^{2x} tends to 0, and so $\tanh x$ will tend to -1. We can rearrange the formula for $\tanh x$ one last time (by dividing top and bottom by e^{2x}), to obtain the expression $\frac{1-e^{-2x}}{1+e^{-2x}}$, and in this form it becomes clear that as x tends to ∞, e^{-2x} tends to 0, and so $\tanh x$ will tend to 1. Combining all of these facts together, we arrive at precisely what we had in Figure 17.4.

18
Motion and Curvature

Test Yourself

If you think you are already comfortable with this material, try these questions first and mark them using the answers at the back of the book. If you get them all right, you're probably ready to move straight on to the next chapter. If some look tricky, study the chapter first and then come back to these when you're ready.

1. A particle of mass m sits on a plane that is inclined at an angle θ to the horizontal. There is no frictional force between the particle and the plane, and the particle is held in place by a constant force of k Newtons, which is applied at an angle of θ *to the plane*. Find (and simplify) an expression for k.

2. A box of mass 1 kg needs to be dragged up a hill, which is inclined at an angle of $45°$ to the horizontal. In order to pull the box, a rope is attached at an angle of $30°$ to the slope, and a constant force of 50 N is to be applied. The coefficient of friction is 0.2. Find the instantaneous acceleration of the box in the first moment that the force is applied.

3. A particle is attached to a string and swung in a horizontal circle of radius 3 m. The angular velocity of the particle is found to be 10 rads^{-1}. What is the acceleration of the particle, and in which direction is this acceleration?

E. Hurst and M. Gould, *Bridging the Gap to University Mathematics*,
DOI: 10.1007/978-1-84800-290-6_18,
© Springer-Verlag London Limited 2009

4. A particle is travelling along a horizontal circular path of radius 2 m with a constant velocity of 8 ms^{-1}. What is the acceleration of the particle?

5. A particle of mass 4 kg is attached to a string and swung in a horizontal circle of radius 2 m. The angular velocity of the particle is found to be 1 rads^{-1}. What force is the particle exerting on the string?

6. A bucket is filled with water, so that its mass is 10 kg, and then tied to a rope which is 2 m long. The bucket is swung in a horizontal circle, using the entire length of the rope, with a constant velocity of 3 ms^{-1}. What is the tension in the rope during this time?

7. Parameterise the curve $y = 3x + 2$.

8. Parameterise the curve $y = \sin x$.

9. Find the length of the curve parameterised by $C = \{(t, 7t + 1) | t \in [0, 3]\}$.

10. By first parameterising the curve, find the length of the curve $y = \cosh x$ between $x = 0$ and $x = \ln 10$.

18.1 Loose Ends or New Beginnings?

This chapter might seem a little bit strange. We're going to start out by taking a look at some more complicated mechanics problems, and we're going to finish by looking at curves. If these two ideas seem totally unrelated to you, don't despair. Before university, "mechanics"-style problems and "pure"-style problems are kept rather distant, but at degree level the two are much more integrated. What a university mathematician would call "applied maths," you would probably consider to be pure; and what a university mathematician would call "pure maths" you would probably consider to be insanity.

Sloping Away

If you've already studied a decent amount of mechanics, you may want to skip this section (and perhaps the next section too, to get straight to the section on curves). If not, then it's time to recall what we did way back in the chapter on mechanics – if you haven't studied this chapter yet, it's time to do so now. We finished the chapter on a bit of a cliffhanger: We saw a little bit of motion in the context of Newton's second law, but then we only looked at *stationary* particles on an inclined plane. Now it's time to unite these concepts and examine some moving particles on a plane.

When the particle was stationary on the plane, we found the magnitude of the frictional force by resolving the particle's weight force into its components parallel and perpendicular to the plane and then making an equality of the form $F_r = mg \sin \theta$. Nothing too mindblowing there. But now consider conducting the following experiment in real life. Take a flat piece of wood and place it on a table. On the piece of wood, place a small item (let's say a sandwich, because as we write this we're feeling rather hungry). Pick up one end of the piece of wood, leaving the other end on the table, and raise this end of the piece of wood so that you have created an inclined plane for the sandwich to sit on. Measure the angle of the incline that you've created and work out how strong the force of friction is. When you've done that, make the plane a little bit steeper and repeat the process. You'll find that the force of friction is a little bit stronger now, because θ has increased a little bit and, in the region of 0-90 degrees, increasing θ increases $\sin \theta$. Then do it again. And again and again and again – you'll find that the force of friction increases each time.

But then you'll go a little bit too far: You'll incline the plane a little more than you did the previous time, and the sandwich will *slip*. You can't work out what the frictional force was in this case using the methods that we used before, because all of them relied on the sandwich being stationary throughout.

As you give up and eat the sandwich, something bothers you. When you made the plane steeper and steeper, the frictional force increased. So *what was* the value of the frictional force? Did it vary?

Yes, it did. This poses us a problem: Given some apparatus, how can we quantify the frictional force in the experiment? Clearly we can't simply state the magnitude of the force because that varies with θ (the angle of the slope). Cunningly, there *is* a useful piece of information that we can extract. It is called "the coefficient of friction" and is given the letter μ. The formula for finding μ is this:

$$F_r = \mu R$$

That is, *at the point of slipping*, the coefficient of friction is equal to the frictional force divided by the normal reaction. Notice the key phrase here: "at the point of slipping." To find the value of μ we need to incline the plane as steeply as we can without having the sandwich slip, so that if we made the slope even the tiniest bit steeper the sandwich would slide down it.

Now that we have this piece of information, we can see how it is useful to us. We saw in the chapter on mechanics that the equation for the normal reaction of the plane on the particle, R, is equal to $mg \cos \theta$. Substituting this into our equation for μ yields the equation:

$$F_r = \mu mg \cos \theta$$

At this point it becomes necessary to summon up Newton's second law again. You'll recall that this is $F = m\frac{dv}{dt} + v\frac{dm}{dt}$, and in this situation we're going to assume that the mass of the particle doesn't vary with time, so that we can use the simplification $F = ma$ (we explored this idea fully in Chapter 10). We look at which direction each force is acting in to determine whether it should be positive or negative in our calculations. Here we'll be trying to find the acceleration of the particle *down* the slope, so we're choosing our "positive direction" to be "down the slope."

The weight force of the particle acting down the slope is $mg\sin\theta$, and the frictional force – which *opposes* motion – we now know to be $\mu mg\cos\theta$. Figure 18.1 is a diagram of this situation, with our forces on.

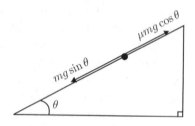

Figure 18.1

Putting all of this together with Newton's second law in its $F = ma$ form gives us the single equation:

$$mg\sin\theta - \mu mg\cos\theta = ma$$

It's really important that you're confident about where all of these things came from. The left-hand side is the force acting down the slope minus the force acting up it, which we can group together and deal with just like they were a single force, in order to equate the whole thing as being equal to ma. The only thing left to do is to cancel the ms, leaving us with the final equation:

$$g\sin\theta - \mu g\cos\theta = a$$

This equation can be used to solve a whole host of problems, so here's a worked example:

> A particle of mass 3 kg is at the point of slipping down a plane inclined at 60° to the horizontal. Given that the coefficient of friction of the system is 0.2, find the particle's initial acceleration at the instant that it is released from rest.

To solve the problem, all that we need to do is to substitute into the equation that was just derived. Here we go:

$$a = g \sin\theta - \mu g \cos\theta$$

$$= \frac{\sqrt{3}}{2} g - 0.2 \frac{g}{2}$$

$$= \left(\frac{\sqrt{3}}{2} - \frac{1}{10} \right) g$$

$$= \left(\frac{5\sqrt{3} - 1}{10} \right) g$$

Now that we've taken a look at motion, we are going to take a look at one final type of question involving particles on inclined surfaces: problems involving external forces.

Outsiders

All of the problems that we've looked at so far (in both this chapter and the chapter on mechanics) involved a particle being left to its own devices on an inclined plane. Nothing other than the particle's weight was pulling it down the plane, and nothing other than friction was opposing this motion. But what happens when we have other forces in the system? How does that affect our calculations?

Thankfully it doesn't make too much difference. There are two key things to remember:

- Resolve *all* forces (including external forces) so that they are expressed in their components acting parallel and perpendicular to the plane.

- Check which direction friction is acting in: friction *always* opposes motion.

The second of these points might seem a little strange at first, but if we have a system where there is strong external force acting *up* the plane, it's very possible that the motion of the particle will be up the slope *and so friction will then be acting in the direction down the slope*: the opposite to what we've been dealing with so far. Remember that this isn't always going to be the case, though: with a steep incline and a small external force acting up the plane, the particle may still well move down the plane, meaning that the direction of the frictional force will still be up it.

Take a look at Figure 18.2 to try to visualise exactly what's going on. (All forces are in Newtons.)

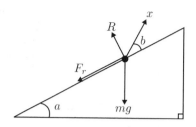

Figure 18.2

You'll see that a force of x Newtons is being applied to the particle on the slope, and that this force is strong enough to mean that the frictional force is acting *down* the slope. Hopefully it's clear that to deal with any problem involving an external force we're going to need to do some resolving of forces, so let's have a think about that. Clearly the standard weight force, mg, will be resolved in the same way as always: $mg \cos a$ will be the pull "into" the plane, and $mg \sin a$ will be the force along the plane. When we resolve the new, external force, however, the sine and the cosine actually come out *the other way around:* this means that for an external force, $x \sin b$ is the force perpendicular to the plane and $x \cos b$ is the force parallel to it. We could spend pages and pages trying to describe this in words, but the only way that you're really going to get to grips with it is if you *do the geometry yourself.* It's really not too scary: have a play around. You know what the result should be now, so once you get that, you know you're on the right track.

There's one last key thing to note before we head to a solution. Remember that the formula for the coefficient of friction was $F_r = \mu R$. This is obviously still true, but when we're finding the value of R we need to remember that R is *only* the force that the plane pushes on the particle with. In situations involving an external force, this won't simply be equal to $mg \sin a$, because now we need to include the fact that the external force is pulling the particle off of the plane a little bit. We'll hopefully clarify this concept a little more in the next example.

Altogether, that's been a lot of theory. Most of the ideas presented here really aren't as bad as they seem, and so hopefully a worked example will be much better at guiding you through than ploughing through even more "explaining."

A particle of mass 2 kg is sitting on a plane inclined at 45° to the horizontal. A string, making an angle of 30° to the plane, pulls on the particle up the plane with a constant force of 20 Newtons. The coefficient of friction of the whole system is 0.6. Find the acceleration of the particle up the plane at the moment that it is released from rest.

Just like with all of these problems, a good diagram is very helpful; take a look at Figure 18.3 now! (All forces are in Newtons.)

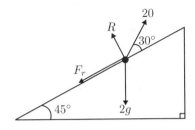

Figure 18.3

First of all, let's take a look at that point that we flagged up just before the question. We need to find the value of R very carefully. Here's the equation that we have in this case:

$$R = 2g \cos 45 - 20 \sin 30$$
$$= \sqrt{2}g - 10$$

Now that we have calculated the correct value for R, we can proceed to find the magnitude of the frictional force:

$$F_r = \mu R$$
$$= 0.6 \left(\frac{2g}{\sqrt{2}} - 10 \right)$$
$$= \frac{1.2g}{\sqrt{2}} - 6$$

Now that we have this calculated, we have *three* forces to use in the final equation: the component of the weight force *and* the frictional force down the plane, and the component of the external force up the plane. To combine them we use $F = ma$ just like before:

$$ma = F$$

$$= \text{Forces } up \text{ the plane } - \text{ Forces } down \text{ the plane}$$

$$2a = 20\cos 30 - 2g\sin 45 - \left(\frac{1.2g}{\sqrt{2}} - 6\right)$$

$$= 10\sqrt{3} + 6 - \frac{3.2g}{\sqrt{2}}$$

$$a = 5\sqrt{3} + 3 - \frac{1.6g}{\sqrt{2}}$$

Make sure you're happy with what happened in that last step: we take the force *up* the plane (because this is the direction that the question requires the answer in), and then subtract *both* the frictional force and the component of the weight force, because they act *down* the plane. Take a good look at the example to make sure you're totally clear about all of the steps, and then tackle these exercises.

EXERCISES

18.1.1. A particle of mass 2 kg is held at rest on a slope that is inclined at an angle of 60° to the horizontal. The coefficient of friction is 0.4. Find the acceleration of the particle at the instant that it is released.

18.1.2. A particle of mass 8 kg is held at rest on a slope that is inclined at an angle of 45° to the horizontal. The coefficient of friction is 0.5. Find the acceleration of the particle at the instant that it is released.

18.1.3. A particle of mass 1 kg is sitting on a slope that is inclined at an angle of 30° to the horizontal. An external force of 100 N pulls the particle up the plane, at an angle of 30° *to the slope*. The coefficient of friction is 0.1. Is the direction of the frictional force in this system up or down the plane?

18.1.4. A particle of mass 100 kg is sitting on a slope that is inclined at an angle of 45° to the horizontal. An external force of 1 N pulls the particle up the plane, at an angle of 30° *to the slope*. The coefficient of friction is 0.6. Is the direction of the frictional force in this system up or down the plane?

18.1.5. A particle of mass 10 kg is held at rest on a slope that is inclined at an angle of 45° to the horizontal. An external force of 20 N acts on the particle, pulling up the plane, at an angle of 30° *to the slope*.

The coefficient of friction is 0.5. Find the instantaneous acceleration of the particle at the moment that it is released from rest.

18.1.6. A particle of mass 10 kg sits on a plane which is inclined at an angle of 30° to the horizontal, and there is no friction between the particle and the plane. An external force acts on the particle, pulling up the plane, at an angle of 30° *to the slope*. Given that the particle is stationary on the slope, find the magnitude of the external force.

18.2 Circular Motion

When it comes to circular motion, most people fall into one of two distinct categories: they're either very experienced and confident in the matter, or they've never really worked with it at all before. If you're confident about circular motion (and therefore about centripetal acceleration), please feel free to skip this section and head straight for the exercises. If you're not, then get ready for some spinning.

Deriving the Tools

In order to get a good idea of what goes on in circular motion, we're going to look at deriving the equations that are used. For a mass travelling at a constant speed, in a *horizontal* circle (vertical movement would involve gravity, and we want to keep things simple!) of radius r, we can think about the motion just like in Figure 18.4.

What the diagram is showing us is that after a small amount of time, Δt, the particle travelling around the circle will have travelled a small distance Δs, and therefore through a small angle, $\Delta \theta$. Now, notice that what we're saying here isn't entirely accurate with the diagram shown. In the diagram it shows that the particle will move along a straight path in order to achieve its distance of Δs. So how do we make this simplification acceptable? Hopefully you'll already be thinking it: We make the time interval so short that Δs is infinitesimally small. This means that we *can* make the approximation with confidence, and that our assertion is valid. Of course, when we are dealing with infinitesimally small time intervals, we're dealing with calculus. Remember also that when we measure in radians, we have arc length $= r\theta$, and when we're taking an infinitesimally small time interval, we can assume that the particle

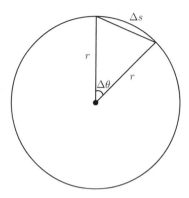

Figure 18.4

is indeed travelling along an arc, so here's what we can state:

$$\Delta s = r \Delta \theta$$

$$\frac{\Delta s}{\Delta t} = r \frac{\Delta \theta}{\Delta t}$$

$$\frac{ds}{dt} = r \frac{d\theta}{dt}$$

If you think back to the standard definitions in mechanics, you'll recall that $\frac{ds}{dt}$ is equal to velocity, because here the s is a displacement. Alongside our standard notion of velocity we also introduce the idea of *angular velocity*. Just as standard velocity is the change in distance over the change in time, angular velocity is the change in angle over the change in time. We use the letter omega, ω, for this and hence:

$$\omega = \frac{d\theta}{dt}$$

Combining all of these factors together, we get the useful expression:

$$v = r\omega$$

We can also derive a useful expression for acceleration in uniform circular motion, but this one requires that we take a little more care, because the arguments are quite subtle. Make sure you keep up!

From Geometry to Swingball

Firstly, consider Figure 18.5.

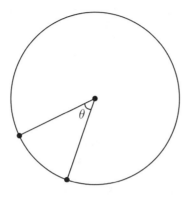

Figure 18.5

The diagram illustrates how, in a short time interval, our particle moves a small distance around the circumference of its circle, "sweeping out" an angle θ. Let's consider the velocity vectors of the particle at the beginning and the end of this short time interval. They look like Figure 18.6.

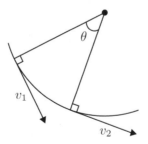

Figure 18.6

An important thing to note is that although the vectors point in slightly different directions, they are of the same magnitude because we know that the particle is travelling at a *uniform* speed (because here we're dealing with *uniform* circular motion). Also, because the motion of the particle is always tangential to the radius of the circle, we know that the velocity vectors must each make a right angle with their respective radii.

Now it's time to think right back to Chapter 5, where we learned how to perform vector subtraction using a diagram. You see, we're interested in finding Δv, which is equal to $v_2 - v_1$. Hopefully you'll recall that all we need to do is to reverse the direction of the arrow for v_1, and place it on the end of v_2: the solution vector will then be the single straight line from the start of v_2 to the end of the reversed v_1, as shown in Figure 18.7.

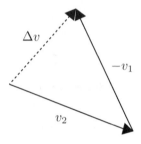

Figure 18.7

You'll now see why we made such a fuss about the velocities acting at right angles to the radii: it allows us to identify angle θ in this diagram – and that's the same angle as the angle that the particle "swept out" during its motion! Look at Figure 18.8 and follow this argument to see why!

- Recall that CD points in the direction of v_1, and DA points in the direction of v_2.

- Consider triangle OAB. We know two of the angles, and so angle $O\hat{B}A$ must be $\frac{\pi}{2} - \theta$, because angles in a triangle sum to π.

- Consider triangle BCD. We now know two of the angles, and so angle $B\hat{D}C$ must be θ.

- Similarly, angle $A\hat{D}E$ must be θ.

In addition to this, we know that both v_1 and v_2 are equal in magnitude and so we can name them both v. The triangle that we have is isosceles, so we can label both of the other angles as α. This gives us Figure 18.9.

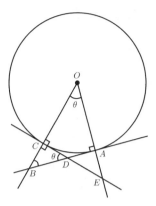

Figure 18.8

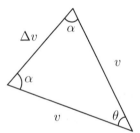

Figure 18.9

We now use the sine rule (which we covered in Chapter 16, if you don't know it from memory) on the triangle to arrive at:

$$\frac{\Delta v}{\sin \theta} = \frac{v}{\sin \alpha}$$

$$\Delta v = \frac{v \sin \theta}{\sin \alpha}$$

Remembering that we're assuming the time interval to be infinitesimally small (so that the particle travels around the edge of the circle, rather than along a chord like our simplified diagram suggests), we can rearrange $v = \frac{\text{arc length}}{\Delta t}$ to arrive at:

$$\Delta t = \frac{\text{arc length}}{v}$$

We're using radians, so the arc length is simply equal to $r\theta$, hence we have:

$$\Delta t = \frac{r\theta}{v}$$

Finally, because $a = \frac{\Delta v}{\Delta t}$ when our time interval is taken to be infinitesimally small, we combine the above expressions to get:

$$a = \frac{\left(\frac{v \sin \theta}{\sin \alpha}\right)}{\left(\frac{r\theta}{v}\right)}$$
$$= \frac{v^2 \sin \theta}{r\theta \sin \alpha}$$

We're almost done! We just need to think carefully about the behaviour of this expression as θ becomes very small. Hopefully you'll have met the small angle approximation $\sin \theta \approx \theta$ when $\theta \approx 0$, so we'll be using that. Also, if we think back to our isosceles triangle, if $\theta \approx 0$ then $\alpha \approx \frac{\pi}{2}$, so $\sin \alpha \approx 1$. Setting these into our expression for a:

$$a = \frac{v^2 \theta}{r\theta \cdot 1}$$
$$= \frac{v^2}{r}$$

And there we have it: an expression for the acceleration. Remembering that we also know $v = r\omega$, we can equivalently write:

$$a = \frac{(r\omega)^2}{r}$$
$$= r\omega^2$$

Making the Right Connections

From these equations, we can again make links with Newton's second law. If we assume that in travelling around a horizontal circular path a particle does not lose any mass, its mass is then not a function of time. Then, again using the reasoning in Chapter 10, we use $F = ma$ to derive the two equations:

$$F = \frac{mv^2}{r}$$
$$F = mr\omega^2$$

Something that might surprise you about all of this is the fact that acceleration must act towards the *centre* of the circle. If you hold a ball on a string and

swing it in a circular path, you might think that it was accelerating radially
outwards from the centre, because you feel a "pull" in the string when you
swing the ball around – but what you're actually feeling is the tension force in
the string that's *causing* the ball to move in a circle in the first place.

All that remains for us to look at is a worked example. From there, the
spotlight is all yours.

A ball of mass 1 kg is tied to a string of length 0.5 m and then swung
around in a horizontal circle at 3 ms^{-1}. Find ω, a and the tension in
the string.

Firstly, let's find ω. Rearranging $v = r\omega$, we see that $\omega = \frac{v}{r}$.

$$\omega = \frac{v}{r} = \frac{3}{0.5} = 6 \text{ rads s}^{-1}$$

Now let's look at the acceleration. This one's just a job for the formula:

$$a = \frac{v^2}{r} = \frac{3^2}{0.5} = 18 \text{ ms}^{-2}$$

Finally, we use the formula that has Newton's Second Law built in to find the
tension:

$$F = \frac{mv^2}{r} = \frac{1 \cdot 3^2}{0.5} = 18 \text{ N}$$

EXERCISES

18.2.1. When a particle is swung in a horizontal circle, in which direction is
it accelerating?

18.2.2. If a particle is undergoing circular motion around a horizontal circle of radius 5 m, and its angular velocity is π rads^{-1}, what is its
velocity?

18.2.3. A particle is travelling at a constant velocity of 4 ms^{-1} around a
horizontal circle of radius 8 m. What is acceleration of the particle?

18.2.4. A particle travelling in a horizontal circle of radius 5 m is measured
to have an angular velocity of 3 rads^{-1}. What is the acceleration of
the particle?

18.2.5. A particle of mass 2 kg is attached to a string and swung in a horizontal circle of radius 2 m, with a constant velocity of 6 ms^{-1}. What is the tension in the string?

18.2.6. A particle of mass 5 kg is attached to a string and swung in a horizontal circle of radius 1 m, with a constant angular velocity of 2 rads^{-1}. What is the tension in the string?

18.3 Curves

As promised in the introduction to the chapter, we're now going to look at some curves. We'll start by exploring a process called the *parameterisation* of curves.

One Line, Many Forms

For the majority of your mathematical career, you'll have been expressing lines in the form $y = mx + c$. The concept of *parameterising* a curve is all about expressing the curve in terms of some parameter which varies between two specified limits (which may be $\pm\infty$).

To help us determine the parameterisation of a straight line, we use vectors. For the vector equation of a straight line, we need to know both a point on the line, $\mathbf{r}_0$, and a direction vector, $\mathbf{d}$. This way, by adding or subtracting multiples of the direction vector from the specified point, we will move along the line that we are defining. This means that the line has been uniquely determined, and hence the vector equation of a straight line (for all real values of t) is:

$$\mathbf{r}(t) = \mathbf{r}_0 + t\mathbf{d}$$

From here, finding the *parameterisation* of the curve is not too much extra work. We need to split the original position vector into its x and y components, so that $\mathbf{r}_0 = (x_0, y_0)$, and then to also do the same to the direction vector, so that $\mathbf{d} = (d_x, d_y)$. To find the parameterisation of the curve, all we do is combine these two facts, and end up with:

$$L = \{(x_0 + td_x, y_o + td_y) | t \in \mathbb{R}\}$$

This is the parameterisation of the line. If you're unsure about any of the notation in this, we covered it in Chapter 11 – it's probably wise to go and revise it now if you're struggling.

Round to Circles

To find the parameterisation of a circle of radius r and centre (m, n), we follow a similar sort of logic: We find an expression in a parameter so as to unambiguously describe the shape that we require. Try to follow the steps in this proof:

- The Cartesian equation for the circle is $(x - m)^2 + (y - n)^2 = r^2$.

- Dividing through by r^2 yields $\frac{(x-m)^2}{r^2} + \frac{(y-n)^2}{r^2} = 1$.

- Using the identity $\sin^2 \theta + \cos^2 \theta = 1$, we can set $\frac{x-m}{r} = \cos t$ and $\frac{y-n}{r} = \sin t$.

- Rearranging each of these, we arrive at $x = m + r \cos t$ and $y = n + r \sin t$.

At this stage, you should be able to spot that the parameterisation of the circle is:

$$C = \{(m + r \cos t, n + r \sin t) | t \in [0, 2\pi)\}$$

Hopefully, everything there will be clear to you. The strange brackets around the 0 and 2π are just there to say to start *at* 0 and finish *just before* 2π (because the point where $t = 2\pi$ would be mapped to the same point as $t = 0$).

Why Do It?

Parameterisation has *many* uses. In your degree, you'll learn loads of them in the first year. Here, we're going to look at just one of them: finding the length of a curve.

If we think about parameterisation visually, what we're actually doing is mapping a part of the real line to a curve in a space. The endpoints of the real line that we deal with are the endpoints of the domain of the parameter, and the "curve in the space" is the curve that we have parameterised. Figure 18.10 is a standard visual explanation of the idea, but if you're not too sure what the diagram is saying (but you are otherwise happy with parameterisation), don't worry about it. Some people find the diagram helpful, but don't let it confuse you if you don't.

Anyway, as promised, we're going to take a look at a use for parameterisation. To find the length of a straight line is easy, but to find the length of a curve that is not straight is actually quite difficult when you think about it. How do you measure it? Make a model using string, and straighten the string out to measure it? Hardly accurate. We can, in fact, use the following formula:

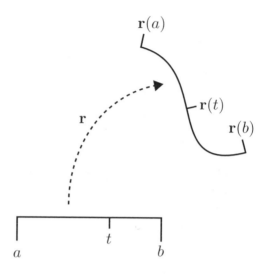

Figure 18.10

$$\text{Length of curve} = \int_a^b \left\| \frac{d\mathbf{r}}{dt} \right\| dt$$

What this formula is doing is looking at the parameterisation of the curve, and seeing how far "along" the end is from the start, as the parameter moves from its lower bound to its upper bound. Note that the **r** there is a *vector*. Here's an example of the formula in action:

> Find the length of the line parameterised by $r(t) = (2 + 3t, 3 + 4t)$, as t varies between 3 and 5.

Firstly, we differentiate $r(t)$. If you've never differentiated a vector with respect to a scalar before, don't worry. You simply differentiate each of the components individually. This gives us:

$$\frac{d\mathbf{r}}{dt} = (3, 4)$$

Therefore, finding the norm of this vector (head on over to the chapter on vectors if you're not sure what this means!) is simply a case of equating:

$$\left\|\frac{d\mathbf{r}}{dt}\right\| = \sqrt{3^2 + 4^2}$$
$$= \sqrt{25}$$
$$= 5$$

This means that all we have to do to find the length of the curve is to solve:

$$\int_3^5 5dt$$

Let's do that now, and then it's time for the final set of exercises in this chapter:

$$\text{Length of curve} = \int_3^5 5dt$$
$$= [5t]_3^5$$
$$= 25 - 15$$
$$= 10$$

EXERCISES

18.3.1. Parameterise the curve $y = x$.

18.3.2. Parameterise the curve $y = x^2$.

18.3.3. Parameterise the circle of radius 2 that is centred on $(3,3)$.

18.3.4. Parameterise the semicircle that starts at $(4,0)$, passes through $(2,2)$ and ends at $(0,0)$.

18.3.5. Consider the curve $C = \{(5t + 8, 12t) | t \in [0,4]\}$. What is the length of C?

18.3.6. Consider the curve $C = \{(8t^2, 6t^2 + 7) | t \in [1,3]\}$. What is the length of C?

18.3.7. What integral would need to be solved to find the length of the curve $y = x^3$ between $x = 0$ and $x = 5$?

18.3.8. What integral would need to be solved to find the length of the curve $y = 2x^2$ between $y = 0$ and $y = 18$?

Where Now?

This chapter is all about new beginnings. In it, you'll have developed your skills learned in the mechanics chapter, so that you're now able to tackle more complicated problems. As ever, we can make the situations that we're modelling more and more complicated, hence making the maths harder and harder.

The work on circular motion really is only an introduction to the topic – along with the idea that many people find surprising: that the acceleration is directed towards the centre of the circle. The questions that were tackled in this chapter are just the tip of the iceberg: how about considering vertical circles, where we have to work *against* gravity on the way up, but then *with it* on the way down? Or swinging a particle that is attached to an elasticated rope? Now that the basic equations are there, it's just a case of making a correct mathematical model of the more complex situations before we can go about solving problems like these.

Parameterisation is an absolutely fundamental skill in the first year of a maths degree. You can use the idea of parameterisation on 3-dimensional surfaces as well as curves, and from there you can calculate lots more interesting things about 3-dimensional shapes. Parameterisation is *definitely* something that you'll be seeing a lot of.

This brings us back to the point on which we started the chapter: Why did we include the seemingly "pure" idea of parameterising a curve into a chapter based on applied mathematics? Imagine we wanted to find the work done (which is equal to the force exerted times the distance) in pushing a box along a curvy path: We use the parameterisation to calculate the length of the curve, just like we explored at the end of the chapter. When we have a parameterisation for our curve, we can also calculate various properties like its *curvature* (intuitively, how much the curve *bends*), which are often vital steps in mechanics problems.

Although written primarily for physicists, *University Physics, 11th Ed.* (H. Young and R. Freedman, Pearson, 2004) is an excellent text for exploring applications of the kind of mathematics that we've seen here. It's absolutely jumbo and has plenty of stuff other than mechanics, but it has some great derivations of the equations we've been using, and goes far beyond the scope of the material in this chapter. It's also an excellent source of examples if you're keen to try some more!

19
Sequences

Test Yourself

If you think you are already comfortable with this material, try these questions first and mark them using the answers at the back of the book. If you get them all right, you're probably ready to move straight on to the next chapter. If some look tricky, study the chapter first and then come back to these when you're ready.

1. Draw the graph of the sequence $a_n = 1, 3, 5, 7, 9, \ldots$.

2. Draw the graph of the sequence $a_n = \frac{n}{2}$.

3. Is the sequence $a_n = 3n + (-1)^n$ monotonic?

4. Does the sequence $a_n = \underbrace{1, 2, 3, 4,}\ \underbrace{2, 3, 4, 5,}\ \underbrace{3, 4, 5, 6,}\ \ldots$ tend to ∞? (The braces are only there to help show how the sequence behaves)

5. Does the sequence $a_n = (-1)^n n$ tend to ∞?

6. Does the sequence $a_n = 2 - \frac{1}{n}$ tend to 0?

7. Does the sequence $a_n = \frac{(-1)^n}{5n}$ tend to 0?

8. Does the sequence $a_n = \frac{3}{10n}$ tend to 0, ∞ or neither?

9. Does the sequence $a_n = (-n)^n$ tend to 0, ∞ or neither?

10. Does the sequence $a_n = \sqrt{n}$ tend to 0, ∞ or neither?

E. Hurst and M. Gould, *Bridging the Gap to University Mathematics*,
DOI: 10.1007/978-1-84800-290-6_19,
© Springer-Verlag London Limited 2009

19.1 (Re)Starting Afresh

At university, one of the important things that you'll do is go right "back to basics," and rigorously *prove* all of the tools that you've been using for years. For example, differentiating is an important part of pre-degree work, but being able to *do* it is a far cry from actually *understanding* what is going on. This sort of investigation, called "analysis," will be a key part of your degree. Your university may not teach analysis in the first year, as some are opting to leave this difficult topic to second years, but still "welcome" their new students with the very basis of analysis: sequences.

What Is a Sequence?

A sequence is simply an *ordered* list of numbers. The ordering is crucial: It is essential that we know which number is in which place in the sequence. Notice too that the definition makes it apparent that a sequence is discrete (we encountered this word when working with inequalities – have a quick recap if you need to). When we plot a sequence on a graph, this means that we get a collection of points and not a continuous line. So the sequence $(a_n) = 1, 2, 3, 4, \ldots$ (which can also be written $a_n = n$) looks like the diagram shown in Figure 19.1.

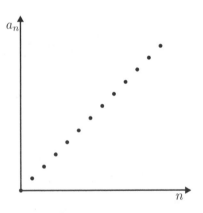

Figure 19.1 $a_n = n$

Bigger, Smaller or the Same Again?

In our study of sequences, it is helpful to be able to identify different properties that we observe. In analysis, all the sequences that are of use to us are infinite – they start with a first term, a_1, but they go on forever. This means that when we identify the following trends, we need them to be true *for all* values in the sequence. In mathematics, we write "for all" so many times that we give it its own symbol: $\forall$. From now on, when we want to say "for all," we will use the $\forall$ symbol instead.

Here are the properties that a sequence may have when we look if it is getting larger or smaller:

A sequence is:

- **Increasing** if $\forall\, n, a_n \leq a_{n+1}$.

- **Strictly Increasing** if $\forall\, n, a_n < a_{n+1}$.

- **Decreasing** if $\forall\, n, a_n \geq a_{n+1}$.

- **Strictly Decreasing** if $\forall\, n, a_n > a_{n+1}$.

Notice the crucial $\forall$ sign. We need our property to be true *everywhere* in the sequence, so it is quite possible that a given sequence will have none of these properties. Also note that by our definitions *it is possible for a sequence to be BOTH increasing and decreasing at the same time.* For example, the sequence $(a_n) = 3, 3, 3, 3, \ldots$ has the property $a_n \leq a_{n+1}$ *and* the property $a_n \geq a_{n+1}$, so although the sequence is neither strictly increasing nor strictly decreasing, it is *both* increasing and decreasing. Did you spot that if a sequence is strictly increasing then it is also increasing? And that if a sequence is strictly decreasing then it is also decreasing?

We have another set of labels to put on sequences, and this time every sequence has either one or the other of these properties. That is, every sequence is either:

- **Monotonic** if it is increasing or decreasing or both.

- **Non-monotonic** if it is neither increasing nor decreasing.

Before we go into the exercises, have one last check that you're up to speed with these definitions, especially in noting that being monotonic **does not** look at whether a sequence is *strictly* increasing or *strictly* decreasing: here, simply increasing or decreasing will do. So many students spend days puzzling over their analysis notes because they never understood these fundamental definitions. Getting them under your belt before you start the course is most definitely time well spent.

EXERCISES

19.1.1. Draw the graph of the sequence $(a_n) = 1, 2, 1, 2, 1, 2, \ldots$.

19.1.2. Draw the graph of the sequence $(a_n) = 2, 4, 6, 8, \ldots$.

19.1.3. Draw the graph of the sequence $(a_n)) = (4 - n)$.

19.1.4. Draw the graph of the sequence $(a_n) = (\sin n)$.

19.1.5. Say which of the following properties each of these sequences have:

- Increasing
- Strictly increasing
- Decreasing
- Strictly decreasing
- Monotonic
- Non-monotonic

a) $(a_n) = (n)$

b) $(a_n) = 10, 10, 9, 9, 8, 8, 7, 7, 6, 6, \ldots$

c) $(a_n) = ((-1)^n)$

d) $(a_n) = (\sin n)$

19.2 To Infinity (Not Beyond)

This concept must be the most efficient "fun-killing" tool at an analysis lecturer's disposal. Over the years, the number of students missing out on partying with their new friends because they had to sit in their rooms, battling it out with this definition, must be in the millions. If there is just one thing that you learn properly from this book before you start your degree, let it be this.

When Sequences Strike Back

Here it is. Like all the nastiest things, it looks so deceptively simple:

A sequence tends to infinity if, after a certain point, its terms are always larger than any number we choose.

Read it a few times and have a think. Cancel all your plans if you need to; we're going to conquer this fiend if it kills us. You see, our "instinctive" notion of what it means to tend to infinity is nowhere near good enough to guide us any more. Look at these two nasty facts, for example:

- Many increasing sequences do not tend to infinity.

- A sequence that tends to infinity doesn't need to be increasing.

Think about why each of these is true, and if you're not sure, review the definition of "increasing" from the previous section. Here's an example to illustrate each:

- $(a_n) = \left(\frac{-1}{n}\right)$. This is definitely an increasing sequence (generate some terms if you don't believe it!), and yet it is never greater than 0, so it can't tend to infinity.

- The sequence $(a_n) = \left(\frac{n}{2} + (-1)^n\right)$ tends to infinity, but it is not increasing (write out some of its terms if you need to convince yourself of this. The sequence isn't monotonic).

Round 2

Now that we've seen some of the problems this nasty definition causes us, let's start to take some steps towards understanding it. For many students, the "Eureka!" moment of suddenly understanding comes from a graphical approach, so we'll try that.

Let's look at that definition again:

A sequence tends to infinity if, after a certain point, its terms are always larger than any number we choose.

Now, to make things easier to visualise, let's deal with "larger than any number we choose" first.

Let's work through the example of $(a_n) = (n)$. We *start* by picking any number: let's choose 5. For our sequence to have any hope of tending to infinity, it must get "larger than any number we choose." Will the sequence $(a_n) = (n)$ get larger than 5? Yes: at $n = 6$, we have what we need. We can verify this property graphically by drawing a horizontal line at the number we "choose," and checking that our sequence gets above it, as shown in Figure 19.2.

Now to verify the more difficult part – that after a certain point, its terms are *always* larger than this. In our sequence $(a_n) = (n)$, this is definitely true.

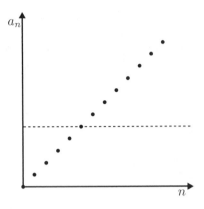

Figure 19.2 $(a_n) = (n)$

But consider the sequence

$$(a_n) = \begin{cases} 0, & n \text{ even} \\ n, & n \text{ odd} \end{cases}$$

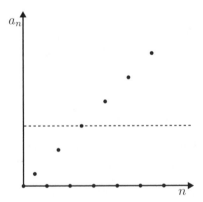

Figure 19.3 $a_n = 0$ for even n, and n for odd n

Figure 19.3 shows what happens if we "choose" 5: Our sequence does get above 5, but then it goes back below it. Our definition requires that there be a point after which the terms are *always* above our chosen number. This won't ever happen for this sequence if we "choose" 5. So we are forced to conclude that our sequence does not tend to infinity. Let's take a look at one more example:

Does the sequence $(a_n) = \underbrace{1, 2, 3}, \underbrace{2, 3, 4}, \underbrace{3, 4, 5}, \underbrace{4, 5, 6}, \dots$ tend to infin-

ity? (The braces are only there to help show how the sequence behaves.)

Let's start by taking a look at a graph of the sequence, as shown in Figure 19.4.

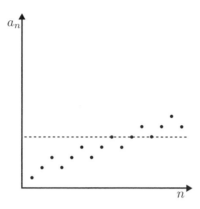

Figure 19.4 $(a_n) = 1, 2, 3, 2, 3, 4, 3, 4, 5, \dots$

Look carefully at what happens. At $n = 12$, $a_n = 6$, so we're above our 5. But at $n = 13$, $a_n = 5$, so we're no longer *above* 5. But wait! At $n = 14$, we're back above 5, and after this point we never go back to the line again (we're always above it). So our sequence *does* meet the criteria when we choose our number to be 5.

The final thing to note is the part of the definition that says "any number we choose." Above, we saw that it works when we chose 5, but would it work for a larger number? In that example, the answer is yes! If we choose 10, 100, 500000,..., we could always find a point in our sequence after which *all* the terms are greater than any number we choose. This means that our sequence does indeed tend to infinity.

It's important to check this "larger than any number we choose" property, and not just assume that a sequence tends to infinity because we find some number for which it works. Look back at the sequence

$$(a_n) = \begin{cases} 0, & n \text{ even} \\ n, & n \text{ odd} \end{cases}$$

Now imagine that we had chosen -1 as our "any number." Certainly all the terms in our sequence are greater than this, so we may be tempted to stop and conclude "tends to infinity." If we choose any positive number as our "any number" instead, we quickly see that this is not the case!

So, the moral of the story, and the final piece of advice before you're out there on your own, is this:

- If you choose a number and there *is not* a point after which every term of the sequence is greater than your number, you can stop and conclude "does not tend to infinity."

- If you choose a number and there *is* a point after which every term of the sequence is greater than your number, **you're not done yet.** You need to check that this holds *whatever* number you choose, not just this specific one i.e. ∀ numbers chosen.

So there you have it. A guide to the nastiest concept in sequences. This section may take a few reads to fully understand, but it'll definitely be worth it in the long run.

EXERCISES

19.2.1. Will the sequence $(a_n) = (2n)$ ever get larger than 12? What, if anything, can we conclude from this fact alone?

19.2.2. Will the sequence $(a_n) = (-n)$ ever get larger than 7? What, if anything, can we conclude from this fact alone?

19.2.3. Is there a point after which every term of the sequence $(a_n) = (n^2)$ is greater than 42? What, if anything, can we conclude from this fact alone?

19.2.4. Is there a point after which every term of the sequence $(a_n) = ((-2)^n)$ is greater than 20? What, if anything, can we conclude from this fact alone?

19.2.5. Does the sequence $(a_n) = (n^2 + n)$ tend to ∞?

19.2.6. Does the sequence $(a_n) = (n + (-1)^n)$ tend to ∞?

19.2.7. Does the sequence $(a_n) = \left(100 - \frac{1}{n}\right)$ tend to ∞?

19.2.8. Does the sequence $(a_n) = \begin{cases} 2n, & n \text{ odd} \\ n, & n \text{ even} \end{cases}$ tend to ∞?

19.3 Nothing at All

After grappling with the "tends to ∞" definition, you hopefully have a good feeling for what it means and what it actually requires. Now that we have it under our belts, we can quickly change a few things round and get another concept mastered: what it means for a sequence to tend to 0.

Crossing the Line

Let's look at some sequences and try to "discover" a definition for tending to 0. The first thing we'll examine is the sequence $(a_n) = \left(\frac{1}{n}\right)$. By drawing a graph, or perhaps by "gut instinct," it seems logical to say that $(a_n) = \left(\frac{1}{n}\right)$ tends to 0. So let's begin our definition by looking at just the behaviour of this sequence: We might guess that the requirement is "a sequence tends to 0 if, after a certain point, its terms are always smaller than any number we choose." But wait – there's a problem. *That's wrong!*

When we were looking at sequences tending to ∞, we wanted the terms to get really large. But when we look at sequences tending to 0, we can't just look for terms getting less and less, because if we're not careful they'll go *way beyond* 0 and down into very negative numbers – perhaps even to $-\infty$! This is clearly not the same as tending to 0.

So how do we deal with this? Let's look now at the sequence $(a_n) = \left(\frac{-1}{n}\right)$. This too tends to 0, but as we go along the terms are actually getting larger!

It's time for our good old friend, the modulus sign, to step up. The modulus of terms in $(a_n) = \left(\frac{1}{n}\right)$ is simply $\frac{1}{n}$, so this sequence is unchanged. But the modulus of terms in $(a_n) = \left(\frac{-1}{n}\right)$ is $\frac{1}{n}$, which we already know tends to 0. So by taking the modulus (often called the "absolute value") of the terms in our sequence, we overcome the problem and arrive at the correct definition:

A sequence tends to 0 if, after a certain point, the absolute value of its terms are always smaller than any positive number we choose.

Note now that we're restricted to choosing positive numbers, because if we chose -1, say, no sequence could ever possibly satisfy our requirements, because of the effect of taking absolute values.

Beyond this, all the concepts work the same way as they did with sequences tending to ∞. Here's a few examples, and then a final set of exercises:

Does the sequence $(a_n) = ((-1)^n)$ tend to 0?

If we take the absolute values, we get the sequence $|a_n| = 1, 1, 1, 1, \ldots$. If we choose the positive number $\frac{1}{2}$, we can instantly see that our sequence does not ever take a value below $\frac{1}{2}$, so cannot possibly tend to 0.

Does the sequence $(a_n) = \left(\frac{(-1)^n}{n}\right)$ tend to 0?

If we take absolute values, we get $|a_n| = 1, \frac{1}{2}, \frac{1}{3}, \frac{1}{4}, \ldots$ (i.e., the sequence $\left(\frac{1}{n}\right)$).
For any positive number we choose, we can always find a point after which every
term in this sequence is less than our chosen number: hence $(a_n) = \left(\frac{(-1)^n}{n}\right)$
does indeed tend to 0.

You're going to need the concept of sequences tending to 0 in order to work
with series in the next chapter, so learn it well!

EXERCISES

19.3.1. Will the sequence $(a_n) = (n)$ ever take values less than 7? What, if
anything, can we conclude from this fact alone?

19.3.2. Will the sequence $(a_n) = (3n)$ ever take values less than 2? What, if
anything, can we conclude from this fact alone?

19.3.3. Is there a point after which every term of the sequence $(a_n) = (5-n)$
is less than 3? What, if anything, can we conclude from this fact
alone?

19.3.4. Is there a point after which the absolute value of every term of the
sequence $(a_n) = ((-3)^n)$ is less than 5? What, if anything, can we
conclude from this fact alone?

19.3.5. Does the sequence $(a_n) = \left(\frac{3}{n}\right)$ tend to 0?

19.3.6. Does the sequence $(a_n) = (10 - n)$ tend to 0?

19.3.7. Does the sequence $(a_n) = \begin{cases} n, & n \text{ odd} \\ 0, & n \text{ even} \end{cases}$ tend to 0?

19.3.8. Does the sequence $(a_n) = \begin{cases} \frac{1}{n}, & n \text{ odd} \\ 0, & n \text{ even} \end{cases}$ tend to 0?

Where Now?

The answer to this question is simply too enormous to write here. From looking
at the terms of sequences, we can look at their sums (these are called series,
and are covered in the next chapter). We can also look at the behaviour of

functions, which are like sequences "with the dots joined up." From there, we can look at differentiation, integration, power series. . . . This well and truly is the tip of the analysis iceberg.

This sort of mathematics will almost certainly be new to you, and many students find getting their heads around analysis very difficult indeed. You'll be in a *much* stronger position early in your degree if you find some time to look over a textbook on analysis: we'd recommend *A Concise Approach to Mathematical Analysis* (M. Robdera, Springer-Verlag, 2003), as it starts right at the beginning. You'll find clear explanations and a good set of exercises to whet your appetite for analysis!

If you've enjoyed learning about logical language in the previous chapters, you might like to learn the *formal* definitions of "tending to infinity" and "tending to 0." These definitions are more precise than the definitions that were given in the chapter (in words), but as a result they look a little scary too! We introduced the "for all" symbol, $\forall$, in this chapter. The only new thing to learn is the symbol $\exists$, which means "there exists." You'll need to be comfortable with the language of sets to be able to understand these definitions properly, so even if you only take the time to get to grips them, without *remembering* them, you'll be setting yourself up well for your analysis studies at university.

In words, our definition for "tends to infinity" was:

A sequence tends to infinity if, after a certain point, its terms are always larger than any number we choose.

Here's the definition in symbols:

$$a_n \to \infty \text{ if } \forall \, \varepsilon > 0, \, \exists \, N \in \mathbb{N} \text{ such that } \forall \, n \geq N, \, a_n > \varepsilon.$$

And we stated "tends to zero" as:

A sequence tends to 0 if, after a certain point, the absolute value of its terms are always smaller than any positive number we choose.

Here's the definition in symbols:

$$a_n \to 0 \text{ if } \forall \, \varepsilon > 0, \, \exists \, N \in \mathbb{N} \text{ such that } \forall \, n \geq N, \, |a_n| < \varepsilon.$$

Test Yourself

If you think you are already comfortable with this material, try these questions
first and mark them using the answers at the back of the book. If you get them
all right, you're probably ready to move straight on to the next chapter. If some
look tricky, study the chapter first and then come back to these when you're
ready.

1. What is the sum of the first n terms of the sequence $(a_n) = (2n)$?

2. Consider $(a_n) = \begin{cases} n+2, & n \text{ odd} \\ 1-n, & n \text{ even} \end{cases}$. Is it possible to find the infinite sum
   of this sequence?

3. Find the infinite sum of the sequence $(a_n) = \begin{cases} \frac{4}{n}, & n \text{ odd} \\ \frac{4}{1-n}, & n \text{ even} \end{cases}$.

4. What property does every sequence with a finite value for its infinite sum
   have?

5. How do we prove that the harmonic sequence does not have a finite value
   for its infinite sum?

6. Use a geometric (i.e., "picture-based") argument to prove that the sequence
   $(a_n) = \left(\frac{1}{3^n}\right)$ has a finite value for its infinite sum.

7. Does the sequence $(a_n) = \left(\frac{7}{n}\right)$ have a finite value for its infinite sum?

E. Hurst and M. Gould, *Bridging the Gap to University Mathematics*,
DOI: 10.1007/978-1-84800-290-6_20,
© Springer-Verlag London Limited 2009

8. Does the sequence $(a_n) = \left(\frac{7}{2^n}\right)$ have a finite value for its infinite sum?

9. Use the comparison test, carefully stating your comparison, to show that $(a_n) = \left(\frac{0.2}{2^n}\right)$ has a finite value for its infinite sum.

10. Use the comparison test, carefully stating your comparison, to show that $(a_n) = \left(\frac{0.3}{n}\right)$ *does not* have a finite value for its infinite sum.

20.1 Various Series

Hopefully you're up to speed on sequences, which was the topic that we covered in the previous chapter. If you're not, then it's going to be *essential* for this chapter, and so it's time to turn to that now. If you feel happy with sequences, then we can take the next step forward: series.

Work on series depends so heavily on work on sequences because *a series is simply the sum of terms in a sequence.* This can be in one of two forms: We can look at finitely many terms or infinitely many terms. For many sequences, finding the sum of finitely many terms is the best that we can do. For example, if you've already studied the chapter on formal logic, you'll be well aware that the sum of the first n terms of the sequence $(a_n) = (n)$ is simply $\frac{n}{2}(n+1)$ (in that chapter, we proved this by induction). Trying to find the sum of infinitely many terms in this series is hopeless: The more terms that we add on, the larger the number gets and the faster that it gets larger! Even the sequence $(a_n) = (1)$ cannot be infinitely summed: the sum of the first n terms is clearly n, but summing infinitely many terms will never yield a finite answer.

So, if even a simple sequence such as $(a_n) = (1)$ proves impossible to infinitely sum, are there *any* sequences out there that we can find the infinite sum of? Well, it may be a boring one, but let's consider the sequence $(a_n) = (0)$. The sum of the first n terms is 0, and we add on 0 at each step. Bingo! As boring as it might be, the infinite sum of the sequence $(a_n) = (0)$ is 0.

Now we've seen *one* sequence whose infinite sum is finite, the logical question to ask is "are there others out there?" *Yes*, there are plenty of them (infinitely many, in fact, but don't think about why this is true until after you've studied the chapter!). The aim of this chapter is to look at some of the different kinds of sequences out there whose infinite sum is finite.

Cancelled

The first type of series that we're going to take a look at is one where many
of the terms "cancel each other out" – that is, there is a negative term for
many of the positive ones (or vice versa, if you prefer). We can "pair up" lots
of the terms and know that the sum of these "cancelling pairs" is simply 0.
That might not be too clear in words, so let's take a look at an example:

Consider the sequence $(a_n) = \begin{cases} n, & n \text{ odd} \\ 1-n, & n \text{ even} \end{cases}$. Is it possible to find
the infinite sum of this sequence?

Now, it might not be too obvious what's going on here at first, so let's generate
some terms in that sequence to see what's happening:

$$(a_n) = 1, -1, 3, -3, 5, -5, \ldots$$

Look what happens if we were to sum up these terms: the sum of the first pair
of terms is 0, the sum of the second pair of terms is 0, the sum of the the third
pair of terms is 0... Sense a pattern here? So can we use this to find the infinite
sum?

Actually, as it turns out, not quite. You see, the problem is that we'll get a
different answer to our sum depending on whether we sum an odd or an even
number of terms. The sum of the first n terms is certainly 0 for even values of
n, but the sum of the first n terms is n for odd values of n. This means that,
annoyingly, we can't find the infinite sum of this sequence – but we're certainly
on the right track to finding some sequences for which we *can*.

Imagine that we had a "cancelling" sequence like the one in the last example,
but where the terms in the sequence tended to 0. Let's take a look at an
example:

Find the infinite sum of the sequence $(a_n) = \begin{cases} \frac{1}{n}, & n \text{ odd} \\ \frac{1}{1-n}, & n \text{ even} \end{cases}$

Again, let's start by generating some terms in this sequence.

$$(a_n) = 1, -1, \frac{1}{3}, -\frac{1}{3}, \frac{1}{5}, -\frac{1}{5}, \frac{1}{7}, -\frac{1}{7}, \ldots$$

Look what's happening: Just like before, the terms are cancelling in pairs. But
as the sequence progresses, the absolute value of the terms is decreasing, and
the sequence is tending to 0. As before, the sum of the first n terms is 0 for
even values of n, but now when we find the *infinite* sum of the sequence, we
know that even if this did involve an odd number of terms, the sum would still
be 0 because *the sequence tends to 0.*

There we have it: we've found another sequence that has a finite answer for its infinite sum. But in doing so, we've actually discovered a very important property of infinite series:

> For a sequence to have a finite sum for infinitely many terms, the sequence *must* tend to 0.

Notice which way around our statement is working (this is a great one for "implies" arrows if you've studied formal logic!) – we're saying that if a sequence has a finite value for its infinite sum, then the sequence *must* tend to 0. We are *not* saying that if a sequence tends to 0 then it must have a finite value for its infinite sum – we'll see a counterexample in the next section. Remember our sequence $(a_n) = (0)$? We saw that the infinite sum of this sequence was finite, and it might seem that a sequence that goes $0, 0, 0, 0, 0 \ldots$ doesn't really *tend* to 0. Strangely enough, if you test that sequence with the definition of "tends to 0" from the previous chapter, this sequence *does* actually tend to 0.

If you're happy with everything that's gone on so far, it's time for some exercises. If you're not (and it's unlikely that you will be after just one read through), then go back for a recap before trying out the questions.

EXERCISES

20.1.1. What is the sum of the first n terms of the sequence $(a_n) = (n^2)$?

20.1.2. What is the sum of the first n terms of the sequence $(a_n) = (2n-1)$?

20.1.3. Is it possible to find the infinite sum of either of the two sequences above?

20.1.4. Consider $(a_n) = \begin{cases} n+1, & n \text{ odd} \\ -n, & n \text{ even} \end{cases}$. Is it possible to find the infinite sum of this sequence?

20.1.5. Find the infinite sum of the following sequence:

$$(a_n) = \begin{cases} \frac{3}{n}, & n \text{ odd} \\ \frac{3}{1-n}, & n \text{ even} \end{cases}$$

20.1.6. Find the infinite sum of the following sequence:

$$(a_n) = \begin{cases} \frac{1}{n}, & n \text{ odd} \\ -\frac{1}{n+1}, & n \text{ even} \end{cases}$$

20.1.7. If we know for certain that a sequence tends to 0, do we know for certain that the infinite sum of the sequence exists?

20.1.8. If we know for certain that the infinite sum of a sequence exists, do we know for certain that the sequence tends to 0?

20.2 Harmonics and Infinities

In the previous section, we saw that all sequences whose infinite sum is a finite number tend to 0. In this section, we're going to look at two different sequences that do indeed tend to 0, and see whether we can find an infinite sum for either of them.

The first sequence that we're going to look at is the sequence $(a_n) = \left(\frac{1}{n}\right)$. This sequence is often known as "the harmonic sequence," and the first few terms are $1, \frac{1}{2}, \frac{1}{3}, \frac{1}{4} \ldots$. Now, this sequence clearly tends to 0 (we saw this in the previous chapter!), but can we find the infinite sum of the terms in it?

As it turns out, the answer is *no*. You see, we can think of the sequence as being lots of small groups of numbers whose sum is greater than or equal to 1. We start a group, and then keep adding subsequent terms into that group until the sum of the group is greater than or equal to $\frac{1}{2}$, at which point we begin a new group and repeat the process. That may not seem very clear, so take a look at some of the terms in the sequence grouped this way:

$$(a_n) = \underbrace{1}_{\geq \frac{1}{2}}, \underbrace{\frac{1}{2}}_{\geq \frac{1}{2}}, \underbrace{\frac{1}{3}, \frac{1}{4}}_{\geq \frac{1}{2}}, \underbrace{\frac{1}{5}, \frac{1}{6}, \frac{1}{7}, \frac{1}{8}}_{\geq \frac{1}{2}} \cdots$$

We can see that each subsequent group is going to need more and more terms in it to make its sum greater than or equal to $\frac{1}{2}$, but we can *always* group the terms in the sequence this way. To clarify the idea, consider the infinite sum of the terms in $a_n = \frac{1}{n}$ and compare them to the terms in another sequence b_n, defined below:

$$a_n = 1, \frac{1}{2}, \frac{1}{3}, \frac{1}{4}, \frac{1}{5}, \frac{1}{6}, \frac{1}{7}, \frac{1}{8} \cdots$$
$$b_n = 1, \frac{1}{2}, \frac{1}{4}, \frac{1}{4}, \frac{1}{8}, \frac{1}{8}, \frac{1}{8}, \frac{1}{8} \cdots$$

It is easy to spot that the terms in a_n are always greater than or equal to the corresponding terms in b_n, but we can cleverly group the terms in b_n so that the sum of the terms in each group (after the first group, which is just the term "1") is *exactly* $\frac{1}{2}$:

$$(b_n) = \underbrace{1}_{=1}, \underbrace{\frac{1}{2}}_{=\frac{1}{2}}, \underbrace{\frac{1}{4}, \frac{1}{4}}_{=\frac{1}{2}}, \underbrace{\frac{1}{8}, \frac{1}{8}, \frac{1}{8}, \frac{1}{8}}_{=\frac{1}{2}} \cdots$$

Therefore we can think of summing the terms in b_n as performing the sum of infinitely many groups, each of which are made up of terms summing to $\frac{1}{2}$. Clearly summing infinitely many halves will give an infinite result, and so the infinite sum of the terms in b_n is infinite. But we observed that the terms in a_n are always greater than or equal to the corresponding terms in b_n, and so the sum of the terms in a_n must be greater than or equal to the sum of the terms in b_n. We can therefore conclude that the infinite sum of the terms in a_n must be infinite too.

So, we've now seen an example of a sequence which *does* tend to 0, but *does not* have a finite value of its infinite sum. Before, the only sequences that we've seen that in fact *do* have a finite value for the infinite sum are the sequence of all 0s and certain sequences where "cancellation" occurs. So are there *any* sequences out there that are made up entirely of strictly positive terms (i.e. banning all negative numbers *and* 0) that have a finite value of the infinite sum?

Let's take a look now at the sequence $(a_n) = \left(\frac{1}{2^n}\right)$. This may look pretty similar to the sequence $(a_n) = \left(\frac{1}{n}\right)$, but clearly the terms in $(a_n) = \left(\frac{1}{2^n}\right)$ get smaller much more quickly. What does this mean for us in our search?

Actually, it means a great deal. Looking at some of the terms in the sequence, we have $(a_n) = \frac{1}{2}, \frac{1}{4}, \frac{1}{8}, \frac{1}{16} \ldots$ And now it's time for pies.

Imagine that every minute, someone gave you some pie, so that at the nth minute you got $\frac{1}{2^n}$th of a pie. That is, in the first minute you got $\frac{1}{2}$ a pie, in the second minute you got $\frac{1}{4}$ of a pie, in the third minute you got $\frac{1}{8}$th of a pie, and so on. Now, rather than scoffing the pie at the earliest opportunity, you decide that you would like to save the pie and put all of the pieces together to see how much you end up with. Figure 20.1 shows what your collection of pie would look like after various stages of time.

Look at what's happening at each stage: the pie is getting nearer and nearer to becoming a whole pie – *but not quite getting there*. Each time, half of the difference between what we have and what we need to make a full pie is filled in. We can repeat this process as many times as we like, but we'll never quite get a single, full pie.

The pie example is simply a way to visualise what's going on numerically in the sequence $(a_n) = \left(\frac{1}{2^n}\right)$. Although the argument needs a little more formality before we can consider it a "rigorous proof," through this crazy example we've observed another sequence that has a finite value for its infinite summation: the sequence $(a_n) = \left(\frac{1}{2^n}\right)$ has the infinite sum of 1. What's more, every term in the sequence is positive, so we have found a strictly positive sequence with a finite value for its infinite sum: exactly the type of sequence that we were looking for. Now, equipped with the adrenaline from such a discovery (and the nutrition of the pie), it's time to tackle some exercises.

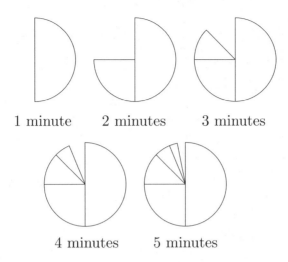

1 minute   2 minutes   3 minutes

4 minutes   5 minutes

Figure 20.1

EXERCISES

20.2.1. What is "The harmonic sequence"?

20.2.2. Does the sequence $(a_n) = \left(\frac{1}{n}\right)$ have a finite value for its infinite sum?

20.2.3. Does the sequence $(a_n) = \left(1 + \frac{1}{n}\right)$ have a finite value for its infinite sum?

20.2.4. Does the sequence $(a_n) = \left(\frac{1}{3^n}\right)$ have a finite value for its infinite sum?

20.2.5. If we use the "pie" analogy again, how much of what we would need to make half a pie would be added every minute in the sequence $(a_n) = \left(\frac{1}{3^n}\right)$?

20.2.6. Does the sequence $(a_n) = \left(\frac{1}{10^n}\right)$ have a finite value for its infinite sum?

20.3 Comparison Testing

One thing that you'll have noticed about the sequences that we've studied in this chapter is that it's often hard to have a "gut instinct" as to whether they will have a finite value as their infinite sum or not. For example, most students (ourselves included) see the harmonic sequence and fully expect a finite value when summing it infinitely – but don't get one! We've used lots of rather different logical arguments in coming to our conclusions about particular sequences, and it may seem at first that working with series can be a bit unpredictable.

Back in the Day

Luckily, many years ago mathematicians felt exactly that, and so set to work finding "tests" that could be applied to sequences, to see whether they would have a finite value as their infinite sum. They came up with two extremely useful results, now known as "the ratio test" and "the comparison test." I'll leave the ratio test to your lecturers, because that one needs a firmer understanding of sequences than we've developed here, but the comparison test is very accessible and *very* helpful.

Proving the comparison test requires a rather deep knowledge of sequences, so this is going to have to be one of those "trust me" times. You'll definitely see a proof early on in your degree studies, but for now, having practice with the test itself is helpful enough to warrant using it without knowing exactly why it works. Anyway, here is the formal statement of the test:

Let $\sum_{n=1}^{\infty} a_n$ and $\sum_{n=1}^{\infty} b_n$ be two series of non-negative real numbers, such that:

- $a_n \leq K \cdot b_n$, for all n and some positive real number K.
- $\sum_{n=1}^{\infty} b_n$ is a finite number.

Then $\sum_{n=1}^{\infty} a_n$ is also finite.

Right, let's decrypt what it is that the comparison test is actually *saying*. We firstly need to have two sequences that we are trying to sum, and neither of those sequences may have any negative numbers in them. Then, if every term in (a_n) is less than or equal to some K times the *corresponding* term in (b_n), *and* we know that we can infinitely sum (b_n) and get a finite result, then we will also get a finite result if we infinitely sum (a_n). Read the formal definition and this explanation a few more times if you need, and make sure that you're happy with what's going on here before you proceed.

Now that we have the test stated, let's start to look at where it might be useful. The first thing to notice is that we need to already know that (b_n) has a finite value for its infinite sum, and so we definitely need to keep in mind all of the sequences that we've studied previously. Let's take a look at an example of the test in action:

Does the sequence $(a_n) = \left(\frac{5}{2^n}\right)$ have a finite value for its infinite sum?

Now, we already know that the sequence $(b_n) = \left(\frac{1}{2^n}\right)$ has a finite value for its infinite sum, and so if we could find a suitable value of K such that $a_n \leq K \cdot b_n$, then we'd have everything we need, because none of the terms in either sequence are negative. Finding such a value of K in this instance really isn't very hard: 5 will do perfectly well, as the inequality that we're dealing with is not strict. So there we have it: even though we don't know what the terms in the sequence $(a_n) = \left(\frac{5}{2^n}\right)$ look like, we know for sure that we can find the infinite sum of this sequence to be a finite value.

Here's one more example before we hit the exercises:

Does the sequence $(a_n) = \left(\frac{5}{n}\right)$ have a finite value for its infinite sum?

To tackle this one, we're going to need to look back at the comparison test and pull a clever trick. The test said that if we could find a value of K such that $a_n \leq K \cdot b_n$ (and all of the other conditions held), then so long as (b_n) has a finite value for its infinite sum, then so does (a_n). But how about if we knew that (b_n) *didn't* have a finite value for its infinite sum? Is there anything helpful we can do there? *Yes.* If we can now find a positive, real value of K such that $a_n > K \cdot b_n$, then we can say for certain that (a_n) *will not have a finite value for its infinite sum.* Follow the logic through: we've just reversed everything. Now, we know already that the sequence $(b_n) = \left(\frac{1}{n}\right)$ does not have a finite value for its infinite sum, and taking K to be $\frac{1}{2}$, for example, will give us the inequality that we need. Hence we can conclude $(a_n) = \left(\frac{5}{n}\right)$ does not have a finite value for its infinite sum, using a sort of "contrapositive version" of the comparison test.

EXERCISES

20.3.1. What condition is there on the terms of both (a_n) and (b_n) in order for us to apply the comparison test?

20.3.2. What conditions are there on the constant K in the comparison test?

20.3.3. Could we use the comparison test (or its "contrapositive version") on the sequence $(a_n) = ((-1)^n)$?

20.3.4. Could we use the comparison test (or its "contrapositive version") on the sequence $(a_n) = \left(3 - \frac{1}{n}\right)$?

20.3.5. Use the comparison test, carefully stating which sequence you are comparing to, to determine whether $(a_n) = \left(\frac{3}{2^n}\right)$ has a finite value for its infinite sum.

20.3.6. Use the comparison test, carefully stating which sequence you are comparing to, to determine whether $(a_n) = \left(1 + \frac{3}{2^n}\right)$ has a finite value for its infinite sum.

Where Now?

As you'll have no doubt noticed, work on series is very much restricted by how much work has already been done on sequences. As your knowledge of sequences increases, you'll be able to use the ratio test in order to determine whether many sequences have a finite value as their infinite sum or not.

From here, you'll be able to look at the two types of convergence: conditional convergence and absolute convergence. Conditional convergence is similar to what we started the chapter with – sequences that have a finite number as their infinite sum only because of the negative terms in them. Absolute convergence is for sequences that have a finite value as their infinite sum *even if we take the absolute values of the terms within them*, hence ignoring any minus signs.

From that point, there is a pretty crazy thing that can be proven: If a sequence is only conditionally convergent and not absolutely convergent, then we can do something special. In theory, changing the order in which we sum terms makes no difference to the result – but as it turns out, changing the order of a conditionally convergent (and not absolutely convergent) series *changes the value of the infinite sum!* If that's not weird enough for you, then you're a very weird individual indeed.

Guide 2 Analysis, 2nd Ed. (M. Hart, Palgrave, 2001) explores many of the clever tricks that you'll learn to make use of when studying analysis at university. It draws on a wide range of tools in the field and it's definitely worth a look before starting your degree.

Appendix

Here is a large set of useful formulae and identities. While you're working through problems, you should be able to find any common identity that you require here. In many university exams, formulae books are not permitted – so when you're busy memorising useful things, this list is a good place to start. Many of the formulae and identities here are beyond the scope of the subject matter of the book itself, so don't worry if there are things that you don't know how to use: this is only meant to be used as a reference.

A.1 Inequalities

The triangle inequality:

- $|x + y| \leq |x| + |y|$
- $|a - c| \leq |a - b| + |b - c|$
- $|x| - |y| \leq |x - y|$
- $|a - b| \geq |(|a| - |b|)| \geq -|a| + |b|$

A.2 Trigonometry, Differentiation and Exponents

- $\frac{d}{dx}(\sin x) = \cos x$

- $\frac{d}{dx}(\cos x) = -\sin x$

- $\frac{d}{dx}(\tan x) = \sec^2 x$

- $\frac{d}{dx}(\sec x) = \sec x \tan x$

- $\frac{d}{dx}(\operatorname{cosec} x) = -\operatorname{cosec} x \cot x$

- $\frac{d}{dx}(\cot x) = -\operatorname{cosec}^2 x$

- $\frac{d}{dx}(\arcsin x) = \frac{1}{\sqrt{1-x^2}}$

- $\frac{d}{dx}(\arccos x) = -\frac{1}{\sqrt{1-x^2}}$

- $\frac{d}{dx}(\arctan x) = \frac{1}{1+x^2}$

- The chain rule: $\frac{d}{dx}(M(x)N(x)) = M'(x)N(x) + M(x)N'(x)$

- The product rule: $\frac{d}{dx}(uv) = v\frac{du}{dx} + u\frac{dv}{dx}$

- The quotient rule: $\frac{d}{dx}\left(\frac{u}{v}\right) = \frac{v\frac{du}{dx} - u\frac{dv}{dx}}{v^2}$

- When $\theta \approx 0$, $\sin\theta \approx \theta$.

- When $\theta \approx 0$, $\cos\theta \approx 1 - \theta$.

- $\log_a b = c \Leftrightarrow a^c = b$

- $\log a + \log b = \log(ab)$

- $\log a - \log b = \log\left(\frac{a}{b}\right)$

- $x\log a = \log(a^x)$

- $\log_b(x) = \frac{\log_k(x)}{\log_k(b)}$

A.3 Polar Coordinates

- $x = r\cos\theta$

- $y = r\sin\theta$

- $x^2 + y^2 = r^2$

- Area of a Sector $= \int \frac{r^2}{2}d\theta$

- Arc length $= \int \sqrt{r^2 + (\frac{dr}{d\theta})^2} d\theta$

- The limaçon: $r = a + b\cos\theta$ or $r = a + b\sin\theta$

- The cardioid: $r = a + b\cos\theta$ or $r = a + b\sin\theta$, and $a = b$

- The lemniscate: $r^2 = a\cos(2\theta)$ or $r^2 = a\sin(2\theta)$

- The Archimedean spiral: $r = a + b\theta$

A.4 Complex Numbers

- $\mathbb{N}$ is the natural numbers, which is the set of strictly positive whole numbers (e.g. 1, 2, 3, ...).

- $\mathbb{Z}$ is the integers, which is the set of all whole numbers (e.g. ...-2, -1, 0, 1, 2, ...)

- $\mathbb{Q}$ is the rational numbers, which is the set $\frac{p}{q}$ where p and q are integers (e.g. $\frac{1}{2}, \frac{7}{5}$).

- $\mathbb{R}$ is the set of real numbers (e.g. 0.84243, ..., $\sqrt{2}$, π).

- $\mathbb{C}$ is the set of complex numbers, which is all the numbers $a + bi$ where a and b are real numbers (e.g. i, $2 + 6i$, $\frac{1}{2} + \sqrt{3}i$).

- $\sqrt{-1} = i$

- $\frac{a+bi}{c+di} = \frac{(a+bi)(c-di)}{c^2+d^2}$

- $(r(\cos\theta + i\sin\theta))^n = r^n(\cos n\theta + i\sin n\theta)$

- $e^{i\theta} = \cos\theta + i\sin\theta$

- n^{th} roots of unity: $x = e^{\frac{2\pi ki}{n}}$

A.5 Vectors

- $\mathbf{u} \cdot \mathbf{v} = |\mathbf{u}||\mathbf{v}|\cos\theta$

- $\mathbf{u} \cdot \mathbf{v} = u_1 v_1 + u_2 v_2 + \cdots + u_n v_n$

- $|\mathbf{u} \times \mathbf{v}| = |\mathbf{u}||\mathbf{v}|\hat{\mathbf{n}}\sin\theta$, where $\hat{\mathbf{n}}$ is a unit vector orthogonal to both $\mathbf{u}$ and $\mathbf{v}$.

$$\bullet \ \mathbf{u} \times \mathbf{v} = \begin{pmatrix} u_2 v_3 - u_3 v_2 \\ u_3 v_1 - u_1 v_3 \\ u_1 v_2 - u_2 v_1 \end{pmatrix} = \begin{vmatrix} \mathbf{i} & \mathbf{j} & \mathbf{k} \\ u_1 & u_2 & u_3 \\ v_1 & v_2 & v_3 \end{vmatrix}$$

- $\mathbf{u} \cdot (\mathbf{v} \times \mathbf{w}) = \mathbf{v} \cdot (\mathbf{w} \times \mathbf{u}) = \mathbf{w} \cdot (\mathbf{u} \times \mathbf{v})$

- $\mathbf{u} \times (\mathbf{v} \times \mathbf{c}) = (\mathbf{u} \cdot \mathbf{w})\mathbf{v} - (\mathbf{u} \cdot \mathbf{v})\mathbf{w}$

- If $\mathbf{u} = u_1\mathbf{i} + u_2\mathbf{j} + u_3\mathbf{k}$ and $\mathbf{v}$ is the direction vector $v_1\mathbf{i} + v_2\mathbf{j} + v_3\mathbf{k}$, then the straight line through $\mathbf{u}$ with direction vector $\mathbf{v}$ has the equation $\frac{x-u_1}{v_1} = \frac{y-u_2}{v_2} = \frac{z-u_3}{v_3}$.

- The plane through position vector $\mathbf{u}$ with normal vector $\mathbf{n} = n_1\mathbf{i} + n_2\mathbf{j} + n_3\mathbf{k}$ has the equation $n_1 x + n_2 y + n_3 z + d = 0$, where $d = -\mathbf{u} \cdot \mathbf{n}$.

- The plane through the point with position vector $\mathbf{u}$ and parallel to $\mathbf{v}$ and $\mathbf{w}$ has equation $\mathbf{r} = \mathbf{u} + s\mathbf{v} + t\mathbf{w}$.

A.6 Matrices

- $\begin{pmatrix} a & b \\ c & d \end{pmatrix} = \begin{pmatrix} d & e \\ f & g \end{pmatrix} \Leftrightarrow a = e, b = f, c = g$ and $d = h$.

- $\begin{pmatrix} a & b \\ c & d \end{pmatrix} + \begin{pmatrix} d & e \\ f & g \end{pmatrix} = \begin{pmatrix} a+d & b+e \\ c+f & d+g \end{pmatrix}$

- $\begin{pmatrix} a & b \\ c & d \end{pmatrix} - \begin{pmatrix} d & e \\ f & g \end{pmatrix} = \begin{pmatrix} a-d & b-e \\ c-f & d-g \end{pmatrix}$

- $\begin{pmatrix} a & b \\ c & d \end{pmatrix} \begin{pmatrix} e & f \\ g & h \end{pmatrix} = \begin{pmatrix} ae+bg & af+bh \\ ce+dg & cf+dh \end{pmatrix}$

- $\begin{vmatrix} a & b \\ c & d \end{vmatrix} = ad - bc$

- $\begin{pmatrix} a & b \\ c & d \end{pmatrix}^{-1} = \frac{1}{ad-bc} \begin{pmatrix} d & -b \\ -c & a \end{pmatrix}$

A.7 Matrices as Maps

- Reflection in the straight line making an angle of θ with the positive x-axis:

$$\begin{pmatrix} \cos(2\alpha) & \sin(2\alpha) \\ \sin(2\alpha) & -\cos(2\alpha) \end{pmatrix}$$

- Anticlockwise rotation by θ about the origin: $\begin{pmatrix} \cos\theta & -\sin\theta \\ \sin\theta & \cos\theta \end{pmatrix}$

- Enlargement by a factor of m horizontally and n vertically: $\begin{pmatrix} m & 0 \\ 0 & n \end{pmatrix}$

- 3D rotation about the x-axis: $\begin{pmatrix} 1 & 0 & 0 \\ 0 & \cos\theta & -\sin\theta \\ 0 & \sin\theta & \cos\theta \end{pmatrix}$

- 3D rotation about the y-axis: $\begin{pmatrix} \cos\theta & 0 & \sin\theta \\ 0 & 1 & 0 \\ -\sin\theta & 0 & \cos\theta \end{pmatrix}$

- 3D rotation about the z-axis: $\begin{pmatrix} \cos\theta & -\sin\theta & 0 \\ \sin\theta & \cos\theta & 0 \\ 0 & 0 & 1 \end{pmatrix}$

A.8 Separable Differential Equations

- $\frac{dy}{dx} P(x)Q(y) = R(x)S(y)$ becomes $\int \frac{Q(y)}{S(y)} dy = \int \frac{R(x)}{P(x)} dx$

- $\int \sin x\, dx = -\cos x + k$

- $\int \cos x\, dx = \sin x + k$

- $\int \tan x\, dx = \ln|\sec x| + k$

- $\int \sec x\, dx = \ln|\sec x + \tan x| + k$

- $\int \mathrm{cosec}\, x\, dx = -\ln|\mathrm{cosec}\, x + \cot x| + k$

- $\int \cot x\, dx = \ln|\sin x| + k$

Partial Fractions:

- $\frac{f(x)}{(x+a)^n} = \frac{A_1}{(x+a)} + \frac{A_2}{(x+a)^2} + \frac{A_3}{(x+a)^3} + \cdots + \frac{A_n}{(x+a)^n}$

- If $px^2 + qx + r$ is an irreducible polynomial: $\frac{f(x)}{(mx+n)(px^2+qx+r)} = \frac{A}{mx+n} + \frac{Bx+C}{px^2+qx+r}$

A.9 Integrating Factors

- $\frac{d}{dx}(P(x)y) = P'(x)y + P(x)\frac{dy}{dx}$

- If $\frac{dy}{dx} + P(x)y = Q(x)$, the integrating factor is $e^{\int P(x)dx}$
- $\frac{dy}{dx} + P(x)y = Q(x)$ becomes $\frac{d}{dx}\left(A(x)B(y)\right) = C(x)$

A.10 Mechanics

- $v = u + at$
- $s = \frac{u+v}{2}t$
- $s = ut + \frac{at^2}{2}$
- $v^2 = u^2 + 2as$

Vector Equations of Motion:

- $\mathbf{v} = \frac{d\mathbf{x}}{dt}$
- $\mathbf{a} = \frac{d\mathbf{v}}{dt}$
- $\mathbf{a} = \mathbf{v}\frac{d\mathbf{v}}{d\mathbf{x}}$

Circular Motion:

- $\mathbf{a}_{rad} = \frac{\mathbf{v}^2}{R}$
- $\omega = \frac{d\theta}{dt}$
- $v = r\omega$
- $a = r\omega^2$
- Tangential velocity: $\mathbf{v} = \omega \times \mathbf{r}$
- Tangential acceleration: $a_{tan} = r\alpha$
- Angular momentum: $\mathbf{L} = \mathbf{r} \times \mathbf{p}$
- Moments of Inertia: $I = \sum_i m_i r_i^2$
- $L = I\omega$
- Rotational kinetic energy: $KE_{\text{rot}} = \frac{I\omega^2}{2}$
- Torque: $\tau = \mathbf{r} \times \mathbf{F}$
- $\tau = \frac{d\mathbf{L}}{dt}$
- For a point mass: $\tau = I\alpha$

Moments:

- Centre of mass $= \frac{m_1\mathbf{r}_1 + m_2\mathbf{r}_2 + \cdots + m_n\mathbf{r}_n}{m_1 + m_2 + \cdots + m_n} = \frac{\sum_i m_i\mathbf{r}_i}{\sum_i m_i}$

- $M\mathbf{R} = \sum_i m_i\mathbf{r}_i$

Rocket Propulsion:

- $\mathbf{F} = \frac{d}{dt}(m\mathbf{v}) = m\frac{d\mathbf{v}}{dt} + v\frac{dm}{dt} = -\mathbf{v}_{ex}\frac{dm}{dt}$

- $\mathbf{a} = \frac{\mathbf{v}_{ex}}{m} \cdot \frac{dm}{dt}$

- $v - v_0 = v_{ex}\ln\frac{m_0}{m}$

Miscellaneous:

- Gravitation: $\mathbf{F} = \frac{-Gm_1m_2}{r^2}\widehat{\mathbf{r}}$

- Gravitational potential energy $= -\frac{Gm_e m}{r}$

- Hooke's law: $F = kx$

- Gyroscope equation: $\Omega = \frac{wr}{I\omega} = \frac{\tau}{L}$

A.11 Logic, Sets and Functions

- Injectivity: $f(a) = f(b) \Rightarrow a = b$ or $a \neq b \Rightarrow f(a) \neq f(b)$

- Surjectivity: If $f : A \to B, \forall\, b \in B\ \exists\, a \in A$ such that $f(a) = b$

- Bijectivity: Both injectivity and surjectivity

Truth Tables:

p	$\Leftrightarrow$	q
T	**T**	T
T	**F**	F
F	**F**	T
F	**T**	F
0	1	0

p	$\Rightarrow$	q
T	**T**	T
T	**F**	F
F	**T**	T
F	**T**	F
0	1	0

•

p	$\wedge$	q
T	**T**	T
T	**F**	F
F	**F**	T
F	**F**	F
0	1	0

•

p	$\vee$	q
T	**T**	T
T	**T**	F
F	**T**	T
F	**F**	F
0	1	0

•

$\neg$	p
F	T
T	F
0	1

Set Language:

Symbol	Meaning
$x \in A$	x is an element of the set A
$y \notin B$	y is not an element of the set B
$M \subset N$	M is a proper subset of N
$P \supset Q$	Q is a proper subset of P
$M \subseteq N$	M is a subset of N
$P \supseteq Q$	P is a subset of Q
$A \cup B$	A or B (or both)
$A \cap B$	A and B
A^C	"Not A"
$A \setminus B$	The complement of A in B
$\emptyset$	"The empty set"
$\{\}$	"The set"
$\{x^2 : x \in \mathbb{Z}\}$	The set of squared xs, such that x is an integer
$\{x^2 | x \in \mathbb{Z}\}$	The set of squared xs, such that x is an integer

A.12 Proof Methods

• Induction: "If it's true for n, then it's true for $n + 1$"

• Strong induction: "If it's true for $1, 2, 3, \ldots, n - 1$, then it's true for n"

- The negated version of $x > y$ is $x \leq y$

- The negated version of $x \geq y$ is $x < y$

- The negated version of $x < y$ is $x \geq y$

- The negated version of $x \leq y$ is $x > y$

A.13 Probability

- $P(A \cup B) = P(A) + P(B) - P(A \cap B)$

- $P(A \cap B) = P(A)P(B|A)$

- $P(A|B) = \frac{P(A \cap B)}{P(B)}$

- Bayes' law: $P(B|A) = \frac{P(A|B)P(B)}{P(A|B)P(B) + P(A|B^C)P(B^C)}$

- $E(X) = \sum_{x_i} x_i p_i$

- $E(aX) = aE(X)$ ("Expectation is a linear operator")

- $\text{Var}(X) = \sigma^2 = E(X - \mu)^2 = \sum_{x_i} x_i^2 p_i - \mu^2$

- $\text{Var}(aX) = a^2 \text{Var}(x)$

- For independent events, $\mathbb{E}(XY) = \mathbb{E}(X)\mathbb{E}(Y)$

- For independent events, $\text{Var}(aX + bY) = a^2 \text{Var}(X) + b^2 \text{Var}(Y)$

- For independent events, $P(A \cap B) = P(A)P(B)$

- $\text{Cov}(X, Y) = \mathbb{E}((X - \mu_X)(Y - \mu_Y)) = \mathbb{E}(XY) - \mu_X \mu_Y$

- Product moment correlation coefficient: $\rho = \frac{\text{Cov}(X,Y)}{\sigma_X \sigma_Y}$

A.14 Distributions

Distribution	$P(X = k)$	Mean	Variance
Binomial(n, p)	$\binom{n}{k} p^k q^{n-k}$	np	npq
Poisson(λ)	$\frac{e^{-\lambda} \lambda^k}{k!}$	λ	λ
Geometric(p)	pq^{x-1}	$\frac{1}{p}$	$\frac{1-p}{p^2}$
Normal(μ, σ^2)	0	μ	σ^2
Uniform	0	$\frac{a+b}{2}$	$\frac{(b-a)^2}{12}$
Exponential	0	$\frac{1}{\lambda}$	$\frac{1}{\lambda^2}$

For Continuous Distributions:

- $\mu = \int x f(x) dx$
- $\text{Var}(X) = \int (X - \mu)^2 f(x) dx$

A.15 Making Decisions

- EMV strategy: Choose $d* \in D$ to maximise $\sum_i d p_i$

A.16 Geometry

- The sine rule: $\frac{a}{\sin A} = \frac{b}{\sin B} \frac{c}{\sin C} = 2R$

- The cosine rule: $a^2 = b^2 + c^2 - 2bc \cos A$

- Pythagoras' theorem: $a^2 = b^2 + c^2$

- Surface area of sphere $= 4\pi r^2$

- Volume of sphere $= \frac{4}{3}\pi r^3$

- A hyperbola has an asymptote at $\frac{x}{a} = \pm\frac{y}{b}$

Conic Sections:

	Ellipse	Parabola	Hyperbola
Standard	$\frac{x^2}{a^2} + \frac{y^2}{b^2} = 1$	$y^2 = 4ax$	$\frac{x^2}{a^2} - \frac{y^2}{b^2} = 1$
Parametric	$(a \cos t, b \sin t)$	$(at^2, 2at)$	$(a \sec t, b \tan t)$ or $(\pm a \cosh t, b \sinh t)$
Eccentricity	$e < 1, b^2 = a^2(1 - e^2)$	1	$e > 1, b^2 = a^2(e^2 - 1)$
Foci	$(\pm ae, 0)$	$(a, 0)$	$(\pm ae, 0)$

A.17 Hyperbolic Trigonometry

- $\cosh x = \frac{e^x + e^{-x}}{2}$
- $\sinh x = \frac{e^x - e^{-x}}{2}$
- $\tanh x = \frac{\sinh x}{\cosh x}$
- $\frac{d}{dx}(\sinh x) = \cosh x$
- $\frac{d}{dx}(\cosh x) = \sinh x$
- $\frac{d}{dx}(\tanh x) = \operatorname{sech}^2 x$
- $\frac{d}{dx}(\operatorname{arsinh} x) = \frac{1}{\sqrt{1+x^2}}$
- $\frac{d}{dx}(\operatorname{arcosh} x) = \frac{1}{\sqrt{x^2-1}}$
- $\frac{d}{dx}(\operatorname{artanh} x) = \frac{1}{1-x^2}$
- $\int \sinh x\, dx = \cosh x$
- $\int \cosh x\, dx = \sinh x$
- $\int \tanh x\, dx = \ln|\cosh x|$
- $\cosh^2 x = 1 + \sinh^2 x$
- $\cosh 2x = \cosh^2 x + \sinh^2 x$
- $\sinh 2x = 2\sinh x \cosh x$

A.18 Motion and Curvature

- Length of curve $= \int_a^b \left\| \frac{d\mathbf{r}}{dt} \right\| dt$
- Unit tangent $\tau = \frac{\frac{d\mathbf{r}}{dt}}{\left\| \frac{d\mathbf{r}}{dt} \right\|}$
- Curvature $\kappa = \frac{\left\| \frac{d\tau}{dt} \right\|}{\left\| \frac{d\mathbf{r}}{dt} \right\|}$

A.19 Sequences

- $\lim_{n\to\infty}(a_n) + \lim_{n\to\infty}(b_n) = \lim_{n\to\infty}(a_n + b_n)$
- $\lim_{n\to\infty}(a_n) \lim_{n\to\infty}(b_n) = \lim_{n\to\infty}(a_n b_n)$

- $\frac{\lim_{n\to\infty}(a_n)}{\lim_{n\to\infty}(b_n)} = \lim_{n\to\infty}\left(\frac{a_n}{b_n}\right)$
- $\lim_{n\to\infty}\left(\frac{1}{n}\right) = 0$
- $\lim_{n\to 0}\left(\frac{1}{n}\right) = \infty$

A.20 Series

- $\sum_{r=1}^{n} r^2 = \frac{n}{6}(n+1)(2n+1)$
- $\sum_{r=1}^{n} r = \frac{n}{2}(n+1)$
- $\sum_{r=1}^{n} r^3 = \frac{n^2}{4}(n+1)^2$
- $s_n = \sum_n a_n$
- If a_n does not tend to 0, s_n does not converge to a finite limit

Taylor's Series:

- $f(x) = f(a) + (x-a)f'(a) + \frac{(x-a)^2}{2!}f''(a) + \cdots + \frac{(x-a)^r}{r!}f^{(r)}(a)$
- $f(a+x) = f(a) + xf'(a) + \frac{x^2}{2!}f''(a) + \cdots + \frac{x^r}{r!}f^{(r)}(a)$
- $\sin x = x - \frac{x^3}{3!} + \frac{x^5}{5!} - \cdots + (-1)^r \frac{x^{2r+1}}{(2r+1)!}$
- $\cos x = 1 - \frac{x^2}{2!} + \frac{x^4}{4!} - \cdots + (-1)^r \frac{x^{2r}}{(2r)!}$
- $\sinh x = x + \frac{x^3}{3!} + \frac{x^5}{5!} + \cdots + \frac{x^{2r+1}}{(2r+1)!}$
- $\cosh x = 1 + \frac{x^2}{2!} + \frac{x^4}{4!} - + \cdots + \frac{x^{2r}}{(2r)!}$
- $e^x = 1 + x + \frac{x^2}{2!} + \cdots + \frac{x^r}{r!}$
- $\ln(x+1) = x - \frac{x^2}{2} + \frac{x^3}{3} - \cdots + (-1)^{r+1}\frac{x^r}{r}$, for $-1 < x \leq 1$

A.21 Pure: Miscellaneous

- $(a+b)^n = a^n + \binom{n}{1}a^{n-1}b + \binom{n}{2}a^{n-2}b^2 + \cdots + \binom{n}{r}a^{n-r}b^r + \cdots + b^n$

- $\binom{n}{r} = {}^nC_r = \frac{n!}{r!(n-r)!}$

- $\dbinom{n}{r} + \dbinom{n}{r+1} = \dbinom{n+1}{r+1}$

- $e^{x \ln a} = a^x$

- $\log_a x = \dfrac{\log_b x}{\log_b a}$

- $\sin(A \pm B) = \sin A \cos B \pm \sin B \cos A$

- $\cos(A \pm B) = \cos A \cos B \mp \sin A \sin B$

- $\tan(A \pm B) = \dfrac{\tan A \pm \tan B}{1 \mp \tan A \tan B}$

- $f'(x) = \lim_{x \to c} \dfrac{f(x) - f(c)}{x - c}$

- $f'(x) = \lim_{h \to 0} \dfrac{f(x+h) - f(x)}{h}$

- The trapezium rule: $\int_a^b y\,dx \approx \dfrac{b-a}{2n}\left((y_0 + y_n) + 2(y_1 + y_2 + \cdots + y_{n-1})\right)$

- The Newton-Raphson process: $f(x) = 0, x_{n+1} = x_n - \dfrac{f(x_n)}{f'(x_n)}$

A.22 Applications: Miscellaneous

- Markov's inequality: $P(X \geq k) \leq \dfrac{E(X)}{k}$

- Chebychev's inequality: $P(|X - \mu| \geq \epsilon) \leq \dfrac{\sigma^2}{\epsilon^2}$

Solid	Moments of Inertia
Slender rod, axis through centre	$\dfrac{ML^2}{12}$
Slender rod, axis through end	$\dfrac{ML^2}{3}$
Rectangular plate, axis through centre	$\dfrac{M(a^2+b^2)}{12}$
Rectangular plate, axis along edge	$\dfrac{Ma^2}{3}$
Hollow cylinder	$\dfrac{M(R_1^2+R_2^2)}{2}$
Thin-walled hollow cylinder	MR^2
Solid cylinder	$\dfrac{MR^2}{2}$
Solid sphere	$\dfrac{2MR^2}{5}$
Thin-walled hollow sphere	$\dfrac{2MR^2}{3}$

Here are some harder questions, many of which tie together concepts from different chapters in the book. They vary quite a lot in difficulty, so don't be disheartened if you find some of them tough. They're meant to see if you can apply some of the concepts explored in the chapters in new (and hopefully interesting!) ways.

1. We say that a matrix A commutes with a matrix B if $AB = BA$. Find what conditions must hold for a matrix $A = \begin{pmatrix} a & b \\ c & d \end{pmatrix}$ to commute with the matrix $B = \begin{pmatrix} 0 & 0 \\ 0 & 1 \end{pmatrix}$.

2. Let $f(x)$ be a quadratic polynomial with real coefficients (i.e., $f(x) = ax^2 + bx + c$ for $a, b, c \in \mathbb{R}$). Show that if a complex number $y = \alpha + \beta i$ is a root of the polynomial, then so is its complex conjugate $\bar{y} = \alpha - \beta i$.

3. Find, in $re^{i\theta}$ form, n points in the complex plane, $\alpha_0, \alpha_1, ..., \alpha_{n-1}$, such that $|\alpha_k| = 1$ for $k = 0, 1, ..., n-1$, $\alpha_0 = 1$ and the αs are evenly spaced on the unit circle with $0 \leq \theta < 2\pi$. We call these the nth roots of unity. Then find an expression for these in the form $x + iy$.

4. Look at the curve shown in Figure B.1. Parameterise this curve in the complex plane.

5. Let
$$A = \begin{pmatrix} 1 & 2 \\ -3 & 6 \end{pmatrix}$$

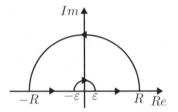

Figure B.1

Find the eigenvalues and then find the two eigenvectors $\mathbf{a} = \begin{pmatrix} a_1 \\ a_2 \end{pmatrix}$ and $\mathbf{b} = \begin{pmatrix} b_1 \\ b_2 \end{pmatrix}$. Next, let the matrix $P = \begin{pmatrix} a_1 & b_1 \\ a_2 & b_2 \end{pmatrix}$ and find P^{-1}. Find the matrix $P^{-1}AP$. What do you notice about this matrix?

6. A sniper aims a rifle at his target from a horizontal distance of 100 m. As he shoots, the bullet leaves the gun with a horizontal velocity of 700 ms^{-1} and the initial vertical velocity is zero. During the bullet's flight, the vertical acceleration is 9.81 ms^{-2} downwards and (ignoring air resistance) the horizontal velocity is constant. Formulate a differential equation for the motion of the bullet and find the vertical distance that the bullet will fall before hitting the target.

7. The police find the sniper's victim at 9:00 am, at which time the temperature of the body is measured to be 28°C. They wait an hour and measure the temperature to be 26°C. The normal temperature of a living body is 37°C and the body is found in a room with temperature 24°C. Newton's law of cooling states:
$$\frac{dT}{dt} = -k(T - A(t))$$
In this equation, T is the temperature of the body, A is the temperature of the surroundings and k is some constant. We take time t in hours. Use the integrating factor method to find an estimate of the time at which the victim was shot.

8. We say that $f : A \to B$ is not injective if there exists $a_1, a_2 \in A$ such that $a_1 \neq a_2$ and $f(a_1) = f(a_2)$. For $g : B \to C$ define the composition $g \circ f(x) = g(f(x))$. Using a contrapositive statement show that if $g \circ f$ is injective then f is also injective.

9. There is a rare blood disease that affects a small number of individuals in the population. An individual has the disease with probability p (where p is very small), and a simple blood test is developed to discover if a sample of blood is infected with disease. It is desired to know exactly which members

of the population have the disease, but performing a test is expensive. In order to help reduce the costs of testing, scientists decide to pool M samples of blood together; that is, they take a blood sample from M individuals, mix all the blood together and perform the test on the mixed sample. If the test from the pooled sample returns a "positive" then every member of the pool is tested individually (to ascertain exactly who is infected), but if the test from the pooled sample returns a "negative," all individuals whose blood was in the pooled sample are free of the disease. Find the value of M which would yield the least expected number of tests to be performed overall. (*Hint:* You may use the fact that, for very small values of p, $(1-p)^M$ is approximately equal to $1 - pM$.)

10. A car is going around a horizontal circular bend of radius 30 m, on a banked, icy road (so there is no friction between the car and the surface) which is inclined at 30 degrees to the horizontal. At what speed should the car travel to stop it from slipping?

11. Show that $8^n - 3^n$ is divisible by 5 $\forall\, n \in \mathbb{N}$. (*Hint:* Use induction!)

12. A sequence (a_n) is called a Cauchy sequence if:

$$\forall\, \varepsilon > 0,\ \exists\, N \in \mathbb{N} \text{ such that } \forall\, n, m > N,\ |a_n - a_m| < \varepsilon$$

Show that the sequence $\left(\frac{1}{n}\right)$ is a Cauchy sequence. (*Hint:* You will need the triangle inequality from the appendix.)

13. Prove that all convergent sequences are Cauchy sequences (*Hint:* You will need the triangle inequality again.)

14. Prove that if a series $\sum_{n=1}^{\infty} a_n$ converges, then the sequence (a_n) converges to 0.

15. Write the contrapositive of the above statement and hence show that $\sum_{n=1}^{\infty} (-1)^n$ does not converge.

16. Prove, by induction, that if both x and y are positive, then $x < y \Rightarrow x^n < y^n$. Then use a contrapositive argument to show that if both x and y are positive, $x^n < y^n \Rightarrow x < y$.

17. You decide to repeatedly flip a biased coin until you flip a head, at which point you will stop and count the total number of times that you flipped the coin. Call this number T. On any single coin flip, the probability of flipping a head is p and the probability of flipping a tail is $q = 1 - p$. Show that, before starting to flip the coin, you should expect to flip the coin $\frac{1}{p}$ times.

18. Given two real numbers a and b, it is always true that: $(a+b)(a-b) = a^2 - b^2$. If we have two matrices A and B whose entries are all real numbers, is it necessarily true that: $(A+B)(A-B) = A^2 - B^2$?

19. Give an example (not already used in the book!) of a sequence that is increasing but that does not tend to ∞. Again, without using one already in the book, give an example of a sequence that tends to ∞ but is not increasing. Finally, give an example of a decreasing sequence in which all the terms are greater than or equal to 0, but that *doesn't* tend to 0.

20. *Formally* prove that if a sequence tends to 0, then it cannot also tend to ∞.

Worked Solutions to Extension Questions

1. We first find:

$$AB = \begin{pmatrix} a & b \\ c & d \end{pmatrix} \begin{pmatrix} 0 & 0 \\ 0 & 1 \end{pmatrix}$$

$$= \begin{pmatrix} 0 & b \\ 0 & d \end{pmatrix}$$

Next we do:

$$BA = \begin{pmatrix} 0 & 0 \\ 0 & 1 \end{pmatrix} \begin{pmatrix} a & b \\ c & d \end{pmatrix}$$

$$= \begin{pmatrix} 0 & 0 \\ c & d \end{pmatrix}$$

Therefore, for the matrices to commute we require:

$$\begin{pmatrix} 0 & b \\ 0 & d \end{pmatrix} = \begin{pmatrix} 0 & 0 \\ c & d \end{pmatrix}$$

So we need $b = 0$ and $c = 0$. Notice that this means that only diagonal matrices commute with the matrix B.

2. Let $y = \alpha + \beta i$. Then:

$$a(\alpha + \beta i)^2 + b(\alpha + \beta i) + c = 0$$
$$a(\alpha^2 + 2\alpha\beta i - \beta^2) + b(\alpha + \beta i) + c = 0$$

Equating coefficients of the real and imaginary parts we have that:

$$a(\alpha^2 - \beta^2) + b\alpha + c = 0$$

$$2a\alpha\beta + b\beta = 0$$

Then for $\bar{y} = \alpha - \beta i$ we have:

$$\begin{aligned} f(\bar{y}) &= a(\alpha - \beta i)^2 + b(\alpha - \beta i) + c \\ &= a(\alpha^2 - 2\alpha\beta i - \beta^2) + b(\alpha - \beta i) + c \\ &= \left[a(\alpha^2 - \beta^2) + b\alpha + c \right] - i\left[2a\alpha\beta + b\beta \right] \end{aligned}$$

But we showed before that both parts in the square brackets are zero, hence $f(\bar{y}) = 0$ so $\bar{y}$ is also a root.

3. Let $\alpha_k = r_k e^{i\theta_k}$. Then $|\alpha_k| = r_k$, so we require that $r_k = 1$. We then need $\alpha_1 = e^{i\theta_1} = 1$ for $0 \le \theta_1 < 2\pi$, so take $\theta_1 = 0$ (as $e^0 = 1$). To be evenly spaced on the unit circle we must have that the α_ks are divided by an angle of $\frac{2\pi}{n}$ (as there are n of them). The first is therefore at an angle $\frac{2\pi}{n}$ from the positive x-axis, the second is at an angle $\frac{2\pi}{n} + \frac{2\pi}{n} = \frac{4\pi}{n}$ from the positive x-axis and so on. Hence the kth root is at an angle $\frac{2k\pi}{n}$ from the positive x-axis. Thus our nth roots of unity are written:

$$\alpha_k = e^{\frac{2\pi k}{n} i}$$

Recall that we make the substitutions:

$$x = r\cos\theta$$

$$y = r\sin\theta$$

Hence using our $re^{i\theta}$ form we have that:

$$x_k = \cos\frac{2\pi k}{n}$$

$$y_k = \sin\frac{2\pi k}{n}$$

So $\alpha_k = x_k + iy_k$ and we obtain the result:

$$\alpha_k = \cos\frac{2\pi k}{n} + i\sin\frac{2\pi k}{n}$$

4. We must divide the curve into four segments. Let γ_1 be the straight line from ε to R, γ_2 be the semicircle from R to $-R$, γ_3 be the straight line from $-R$ to $-\varepsilon$ and let γ_4 be the semicircle from $-\varepsilon$ to ε.

To find γ_1 we need a to find a line that starts at ε and travels a distance $R - \varepsilon$. So we parameterise the curve by:

$$\gamma_1 = \{\varepsilon + (R - \varepsilon)t \, : \, t \in [0, 1]\}$$

Similarly, to find γ_3 we need a to find a line that starts at $-R$ and travels a distance $R - \varepsilon$. So we parameterise the curve by:

$$\gamma_3 = \{-R + (R - \varepsilon)t \, : \, t \in [0, 1]\}$$

To find γ_2 we must think of the semicircle as a curve that keeps a constant distance from the origin while the angle from the x-axis varies. Hence in the $re^{i\theta}$ form we must have $r = R$ and θ varies between 0 and π. So we parameterise the curve by:

$$\gamma_2 = \left\{ Re^{it} \, : \, t \in [0, \pi] \right\}$$

Similarly, to find γ_4 in the $re^{i\theta}$ form we must have $r = \varepsilon$ and θ varies between π and 0 (that is, from π down to 0). So we parameterise the curve by:

$$\gamma_4 = \left\{ \varepsilon e^{i(\pi-t)} \, : \, t \in [0, \pi] \right\}$$

The $(\pi - t)$ involved in the expression for γ_4 comes from the fact that we want the curve to be "swept out" *clockwise*. If you're unsure of why this works, try substituting in some values of t to see what you get.

5. Recall that we find eigenvalues λ from the equation $\det(A - \lambda I_2) = 0$.

$$(A - \lambda I_2) = \begin{pmatrix} 1 - \lambda & 2 \\ -3 & 6 - \lambda \end{pmatrix}$$

So we have:

$$\begin{aligned}
\det(A - \lambda I_2) &= (1 - \lambda)(6 - \lambda) + 6 \\
&= \lambda^2 - 7\lambda + 6 + 6 \\
&= \lambda^2 - 7\lambda + 12 \\
&= (\lambda - 3)(\lambda - 4) = 0
\end{aligned}$$

So our eigenvalues are 3 and 4. For general λ the eigenvectors $\mathbf{v}$ must satisfy $A\mathbf{v} = \lambda\mathbf{v}$, or equivalently $A\mathbf{v} - \lambda\mathbf{v} = (A - \lambda I_2)\mathbf{v} = 0$. So for $\lambda = 3$ we require:

$$\begin{pmatrix} 1 - 3 & 2 \\ -3 & 6 - 3 \end{pmatrix} \begin{pmatrix} a_1 \\ a_2 \end{pmatrix} = \begin{pmatrix} -2 & 2 \\ -3 & 3 \end{pmatrix} \begin{pmatrix} a_1 \\ a_2 \end{pmatrix} = 0$$

Hence $\begin{pmatrix} a_1 \\ a_2 \end{pmatrix} = \begin{pmatrix} 1 \\ 1 \end{pmatrix}$ is a corresponding eigenvector. For $\lambda = 4$ we require:

$$\begin{pmatrix} 1-4 & 2 \\ -3 & 6-4 \end{pmatrix} \begin{pmatrix} a_1 \\ a_2 \end{pmatrix} = \begin{pmatrix} -3 & 2 \\ -3 & 2 \end{pmatrix} \begin{pmatrix} a_1 \\ a_2 \end{pmatrix} = 0$$

Hence $\begin{pmatrix} a_1 \\ a_2 \end{pmatrix} = \begin{pmatrix} 2 \\ 3 \end{pmatrix}$ is a corresponding eigenvector. Therefore:

$$P = \begin{pmatrix} 1 & 2 \\ 1 & 3 \end{pmatrix}$$

The determinant of P is $3 - 2 = 1$. Hence we have that:

$$P^{-1} = \begin{pmatrix} 3 & -2 \\ -1 & 1 \end{pmatrix}$$

Then we can find:

$$P^{-1}AP = \begin{pmatrix} 3 & -2 \\ -1 & 1 \end{pmatrix} \begin{pmatrix} 1 & 2 \\ -3 & 6 \end{pmatrix} \begin{pmatrix} 1 & 2 \\ 1 & 3 \end{pmatrix}$$

$$= \begin{pmatrix} 3 & -2 \\ -1 & 1 \end{pmatrix} \begin{pmatrix} 3 & 8 \\ 3 & 12 \end{pmatrix}$$

$$= \begin{pmatrix} 3 & 0 \\ 0 & 4 \end{pmatrix}$$

Notice: we have a diagonal matrix whose entries are the eigenvalues of A.

6. We have constant horizontal velocity, so to find the time the bullet takes to hit the target we do:

$$\frac{100}{700} = \frac{1}{7}\text{s}$$

Then acceleration $a = \frac{dv}{dt}$ where v is velocity, hence:

$$\frac{dv}{dt} = 9.81$$

So we can separate the variables and integrate:

$$\int_0^t dv = \int_0^t 9.81 \ dt$$

$$v(t) = 9.81t + c$$

But the initial vertical velocity is 0, so when $t = 0$ we have $v = 0$. So the constant c is 0. Then we use the fact that velocity $v = \frac{dx}{dt}$ where x is displacement, and so:

$$\frac{dx}{dt} = 9.81t$$

Once again we can separate the variables and integrate:

$$\int_0^t dx = \int_0^t 9.81t \; dt$$

$$x(t) = \frac{1}{2}9.81t^2 + k$$

We can set $x = 0$ when $t = 0$, hence k is 0. We then use the fact that the bullet travels for $\frac{1}{7}$s to see:

$$x = \frac{1}{2} \cdot 9.81 \left(\frac{1}{7}\right)^2$$

$$= \frac{1}{2} \cdot 9.81 \left(\frac{1}{49}\right)$$

$$= 0.100...\text{m}$$

So the bullet will fall approximately 10 cm before hitting the target.

7. Rearranging the equation we get:

$$\frac{dT}{dt} + kt = kA(t)$$

From the integrating factor method we must then multiply through by $e^{\int k \; dt} = e^{kt}$, hence we have:

$$\frac{dT}{dt}e^{kt} + kte^{kt} = kA(t)e^{kt}$$

We can then write this as:

$$\frac{d}{dt}\left(Te^{kt}\right) = kA(t)e^{kt}$$

Integrating both sides between the limits t_1 and t_2 we have:

$$\int_{t_1}^{t_2} \frac{d}{dt}\left(Te^{kt}\right) \; dt = \int_{t_1}^{t_2} kA(t)e^{kt} \; dt$$

$$T(t_2)e^{kt_2} - T(t_1)e^{kt_1} = A(t)e^{kt_2} - A(t)e^{kt_1}$$

To find k we must enter some values. We have that $A = 24$, $t_2 = 10$, $t_1 = 9$, $T(t_2) = 26$ and $T(t_1) = 28$. Hence:

$$26e^{10k} - 28e^{9k} = 24e^{10k} - 24e^{9k}$$

$$26e^k - 28 = 24e^k - 24$$

$$26e^k - 24e^k = 28 - 24$$

$$2e^k = 4$$

$$e^k = 2$$

$$k = \ln 2$$

Then we go back to the equation, substituting in the values $A = 24$, $t_1 = 9$, $T(t_1) = 28$, $T(t_0) = 37$ and k as above. This gives us:

$$28e^{9\ln 2} - 37e^{t_0 \ln 2} = 24e^{9\ln 2} - 24e^{t_0 \ln 2}$$

$$28e^{\ln 2^9} - 37e^{\ln 2^{t_0}} = 24e^{\ln 2^9} - 24e^{\ln 2^{t_0}}$$

$$28e^{\ln 2^9} - 24e^{\ln 2^9} = 37e^{\ln 2^{t_0}} - 24e^{\ln 2^{t_0}}$$

$$4e^{\ln 2^9} = 13e^{\ln 2^{t_0}}$$

$$\frac{4e^{\ln 2^9}}{13} = e^{t_0 \ln 2}$$

$$\ln\left(\frac{4e^{\ln 2^9}}{13}\right) = t_0 \ln 2$$

Hence we have that:

$$t_0 = \frac{\ln\left(\frac{4e^{\ln 2^9}}{13}\right)}{\ln 2} = 7.299... \approx 7.30$$

So the approximated time of death is 7:18 am. (Remember, it's 0.299 hours.)

8. If f is not injective then there exist $a_1, a_2 \in A$ such that $a_1 \neq a_2$ and $f(a_1) = f(a_2)$. Therefore, this implies that $g(f(a_1)) = g(f(a_2))$. So we have that there exist $a_1, a_2 \in A$ such that $a_1 \neq a_2$ and $g(f(a_1)) = g(f(a_2))$, so $g \circ f$ is not injective. By the contrapositive argument we have that if $g \circ f$ is injective then f is also injective.

9. Let N be the number of tests that we perform on a block of M people. Clearly either $N = 1$ (if the pooled test returns "negative") or $N = M + 1$ (if the pooled test returns "positive," in which case there is one pooled test and M individual tests). So:

$$\mathbb{E}(N) = 1 \cdot \mathbb{P}(N = 1) + (M + 1)\mathbb{P}(N = M + 1)$$

The probability of an individual *not* having the disease is $1 - p$. For a pooled sample from M individuals to return "negative," all M individuals must *not* have the disease, and so the probability of this happening is $(1 - p)^M$. So:

$$\mathbb{E}(N) = (1 - p)^M + (M + 1)(1 - (1 - p)^M)$$
$$= M + 1 - M(1 - p)^M$$

So the expected number of tests per person is:

$$\frac{\mathbb{E}(N)}{M} = 1 + \frac{1}{M} - (1 - p)^M$$

Use the approximation given in the question:

$$\frac{\mathbb{E}(N)}{M} = 1 + \frac{1}{M} - (1 - pM)$$
$$= \frac{1}{M} + pM$$

We need to minimise $\frac{\mathbb{E}(N)}{M}$. We can do this by differentiating:

$$\frac{d}{dM}\left(\frac{\mathbb{E}(N)}{M}\right) = -\frac{1}{M^2} + p$$

To find the minimum, set this equal to 0 and solve:

$$\frac{1}{M^2} = p$$
$$M^2 = \frac{1}{p}$$
$$M = \frac{1}{\sqrt{p}}$$

We take the positive square root because having a negative value for M doesn't make sense!

10. We will call the mass of the car m and the contact force R. Then, resolving vertically we have that:

$$R\cos 30 = mg$$
$$R = \frac{2mg}{\sqrt{3}}$$

Notice that this is greater than the weight of the car, because the car's circular motion causes it to be forced "into" the road, and hence the reaction force increases.

We then resolve horizontally, using the equation for circular motion $F = \frac{mv^2}{r}$ with $r = 30$ m:

$$R\sin 30 = \frac{mv^2}{30}$$
$$v^2 = \frac{15\left(\frac{2mg}{\sqrt{3}}\right)}{m}$$
$$= 10\sqrt{3}g$$
$$v = \sqrt{10\sqrt{3}g} \text{ ms}^{-1}$$

11. For $n = 1$ we have $8 - 3 = 5$ which is divisible by 5. Then, if it is true for the nth case, for $n + 1$ we have:

$$8^{n+1} - 3^{n+1} = 8^{n+1} - 3 \cdot 8^n + 3 \cdot 8^n - 3^{n+1}$$
$$= 8^n(8 - 3) + 3(8^n - 3^n)$$
$$= 8^n \cdot 5 + 3(8^n - 3^n)$$

It is clear that the $8^n \cdot 5$ is always divisible by 5 and $8^n - 3^n$ is the nth case. Hence, by induction, this holds for all n.

12. We have that $(a_n) = \left(\frac{1}{n}\right)$ and $(a_m) = \left(\frac{1}{m}\right)$. So for the sequence to be Cauchy we require:

$$\left| \frac{1}{n} - \frac{1}{m} \right| < \varepsilon$$

Using the triangle inequality, we have that:

$$\left| \frac{1}{n} - \frac{1}{m} \right| \leq \left| \frac{1}{n} \right| + \left| \frac{1}{m} \right|$$

Therefore if we can choose an N such that $\frac{1}{n} < \frac{\varepsilon}{2}$ and $\frac{1}{m} < \frac{\varepsilon}{2}$ then the definition will hold. Rearranging, we require $m, n > \frac{2}{\varepsilon}$. So if we choose $N > \frac{2}{\varepsilon}$ we have that $\forall \, \varepsilon > 0, \, \exists \, N \in \mathbb{N}, N > \frac{2}{\varepsilon}$ such that $\forall \, n, m > N$:

$$\left| \frac{1}{n} - \frac{1}{m} \right| \leq \left| \frac{1}{n} \right| + \left| \frac{1}{m} \right|$$
$$< \frac{\varepsilon}{2} + \frac{\varepsilon}{2} = \varepsilon$$

13. If a sequence (a_n) converges to a limit a then:

$$\forall \, \frac{\varepsilon}{2} > 0, \, \exists \, N_1 \in \mathbb{N} \text{ such that } \forall \, n > N_1, \, |a_n - a| < \frac{\varepsilon}{2}$$

Similarly we can say that:

$$\forall \, \frac{\varepsilon}{2} > 0, \, \exists \, N_2 \in \mathbb{N} \text{ such that } \forall \, m > N_2, \, |a_m - a| < \frac{\varepsilon}{2}$$

(Note that $\varepsilon > 0 \Leftrightarrow \frac{\varepsilon}{2} > 0$). Then if we pick $N = \max\{N_1, N_2\}$, both of these statements hold. Hence $\forall \, \varepsilon > 0 \, \exists \, N \in \mathbb{N}$ such that $\forall \, n, m > N$:

$$|a_n - a_m| = |a_n - a + a - a_m|$$
$$\leq |a_n - a| + |a - a_m|$$
$$= |a_n - a| + |a_m - a|$$
$$< \frac{\varepsilon}{2} + \frac{\varepsilon}{2} = \varepsilon$$

The top line is by the triangle inequality. Also note that the third line holds because $|x| = |-x|$ for any x.

This is precisely the definition of what it means for a sequence to be Cauchy.

14. If a series $\sum_{n=1}^{\infty} a_n$ converges then the partial sums $\sum_{n=1}^{n} a_n$ converge. Hence $\left(\sum_{n=1}^{n} a_n - \sum_{n=1}^{n-1} a_n\right)$ tends to 0. But this is simply:

$$(a_0 + \cdots + a_{n-1} + a_n) - (a_0 + \cdots + a_{n-1}) = a_n$$

Hence a_n tends to 0.

15. The contrapositive is "if the sequence (a_n) does not converge to 0 then $\sum_{n=1}^{\infty} a_n$ does not converge." It is clear that the sequence $((-1)^n)$ does not converge to 0, therefore $\sum_{n=1}^{\infty} (-1)^n$ does not converge.

16. For $n = 1$ we have that $x < y \Rightarrow x < y$, which is obviously true. Multiplying through by x^k yields:

$$x \cdot x^k < y \cdot x^k$$

Then if also $x^k < y^k$, we can note:

$$x \cdot x^k < y \cdot x^k < y \cdot y^k$$

Therefore, we have:

$$x^{k+1} < y^{k+1}$$

So the first statement holds by induction.

The contrapositive of the the second statement is: "$x \geq y \Rightarrow x^n \geq y^n$." For $n = 1$ we have that $x \geq y \Rightarrow x \geq y$ which holds. If we also have that $x^k \geq y^k$ then we can again arrive at:

$$x \cdot x^k \geq y \cdot y^k$$

So we get:

$$x^{k+1} \geq y^{k+1}$$

So, by induction, the statement is true.

17. For T to equal some $n \in \mathbb{N}$, we would need a string of $n - 1$ tails followed by a head. Each flip of the coin is independent, so the probability of this happening is $q^{n-1}p$. We can therefore write $\mathbb{P}(T = n) = q^{n-1}p$. Then the expected number of flips is:

$$\mathbb{E}(T) = \sum_{n=1}^{\infty} n\mathbb{P}(T = n)$$

$$= \sum_{n=1}^{\infty} nq^{n-1}p$$

$$= p(1 + 2q + 3q^2 + \cdots)$$

Notice that $1 + q + q^2 + q^3 + \cdots$ is an infinite geometric series, whose sum is $\frac{1}{1-q}$. So:

$$1 + q + q^2 + \cdots = \frac{1}{1-q}$$

If we differentiate both sides with respect to q:

$$1 + 2q + 3q^2 + \cdots = \frac{1}{(1-q)^2}$$

Plugging this back into the expression for the expectation yields:

$$p\left(\frac{1}{(1-q)^2}\right)$$

Recalling that $1 - q = p$ gives the final answer $\frac{p}{p^2} = \frac{1}{p}$.

18. No, the statement isn't true. The reason that this works for real numbers is that real numbers have the property that $ab = ba$, so:

$$\begin{aligned}(a+b)(a-b) &= a^2 - ab + ba - b^2 \\ &= a^2 - ab + ab - b^2 \\ &= a^2 - b^2\end{aligned}$$

But for matrices A and B it is *not* always true that $AB = BA$ (recall extension question 1!), so we can't say that $-AB + BA = 0$.

19. There are many examples of each of the types of sequences required. Some examples for the first type are $(a_n) = \left(\frac{-1}{n}\right)$ or $(a_n) = 1, 1, 1, 1, 1, \ldots$, of the second type are $(a_n) = 1, 2, 3, 2, 3, 4, 3, 4, 5, 4, 5, 6, \ldots$ or $(a_n) = (n + \sin n)$. Finally, for the third type any sequence of the form $(a_n) = \left(k + \frac{1}{n}\right)$, where k is a positive real number, will do.

20. Writing the definition of "$(a_n) \to 0$" in formal language gives:

$$(a_n) \to 0 \Leftrightarrow \forall\, \varepsilon > 0 \; \exists\, N_1 \in \mathbb{N} \text{ such that } |a_n| < \varepsilon \; \forall\, n \geq N_1$$

Similarly, the definition of "$(a_n) \to \infty$" is, in formal language:

$$(a_n) \to \infty \Leftrightarrow \forall\, C > 0 \; \exists\, N_2 \in \mathbb{N} \text{ such that } a_n > C \; \forall\, n \geq N_2$$

Using the first definition, fix an $\varepsilon > 0$ and find the corresponding N_1. Then choose any $C > \varepsilon$. We know that $\forall\, n \geq N_1$, we must have $|a_n| < \varepsilon$, and so $|a_n| < C$ and so $a_n < C$. Hence no such N_2 could ever exist and hence (a_n) cannot tend to ∞.

Solutions to Exercises

Chapter 1

Test Yourself (page 1):

1. 5,6,7.
2. $x < 5$
3. $x \leq 1$
4. $x > 4$
5. $x \geq 1$
6. $x < -1$ and $x > 1$
7. $-1 < x < 4$
8. $x < 0$ and $x > 3$
9. $-10 < x < 4$
10. $\frac{-3-\sqrt{69}}{2} < x < 2$ and $x > \frac{-3+\sqrt{69}}{2}$

Exercise 1.1 (page 4):

1. 8,9,10.
2. $x > 11$
3. $x > 7$
4. $x \leq 5$
5. $x < 5$
6. $x \leq -10$
7. $x < -12$
8. $x \leq 2$
9. $x > \frac{-2}{3}$
10. $x > 1$

Exercise 1.2 (page 6):

1. $x < 5$
2. $x \leq -3$ and $x \geq 3$
3. $-10 < x < 10$
4. $x < -4$ and $x > 4$
5. No solutions
6. $x < \frac{-4}{3}$
7. $0 < x < 4$
8. $x \leq 0$ and $x \geq 9$

Exercise 1.3 (page 11):

1. $x < -2$ and $x > 2$
2. $x < 0$ and $x > 2$
3. $0 < x < 7$
4. $x < -3$ and $x > -1$
5. $-3 - \sqrt{12} < x < -3 + \sqrt{12}$
6. $x < -8$ and $x > -4$
7. $-6 < x < 3$
8. $x < -5$ and $2 < x < 3$
9. $x < \frac{11}{4}$ and $x > 3$
10. $2 - \sqrt{12} < x < 3$ and $x > 2 + \sqrt{12}$

Chapter 2

Test Yourself (page 13):

1. Divide through by $\sin^2 x$ to arrive at $1 + \cot^2 x = \text{cosec}^2 x$, and then rearrange.

2. $\frac{\sqrt{2}+\sqrt{6}}{4}$

3. $-\frac{\sqrt{3}}{2}$

4. $\tan(2x) = \frac{\sin(2x)}{\cos(2x)} = \frac{2\sin x \cos x}{1-2\sin^2 x}$. Then divide top and bottom by $\cos^2 x$.

5. $2x \, \text{cosec} \, x - x^2 \cot x \, \text{cosec} \, x$

6. $-4\sin(2x)\cos(2x)$

7. $\frac{\sqrt{3}}{8}$

8. e^{10x}

9. $4e^{2x} + 8xe^{2x}$

10. $\frac{2}{x}$

Exercise 2.1 (page 19):

1. LHS $= \frac{\sin^3 x}{\cos x} + \sin x \cos x = \frac{\sin x}{\cos x} - \frac{\sin x \cos^2 x}{\cos x} + \sin x \cos x = \tan x$.

2. LHS $= \frac{\sin^2 x}{\cos x} = \frac{1-\cos^2 x}{\cos x} = \sec x - \cos x$

3. $\frac{\sqrt{6}-\sqrt{2}}{4}$

4. $\frac{\sqrt{6}-\sqrt{2}}{4}$

5. $\frac{1}{2}$

6. $\frac{\sqrt{2}-\sqrt{6}}{4}$

7. $\sin^2 x = \frac{1}{2}\left(1 - \cos(2x)\right)$

8. LHS $= 2\sin x \cos x \frac{\sin x}{\cos x} = 2\sin^2 x = 1 - \cos(2x)$

9. LHS $= \frac{\sin(2x)}{2\cos^2 x}(2\cos^2 x - 1) = \sin(2x) - \frac{\sin(2x)}{2\cos^2 x} = \sin(2x) - \tan x$

10. $\tan(A \pm B) = \frac{\sin(A\pm B)}{\cos(A\pm B)} = \frac{\sin A \cos B \pm \sin B \cos A}{\cos A \cos B \mp \sin A \sin B}$. Then divide top and bottom by $\cos A \cos B$ and cancel.

Exercise 2.2 (page 23):

1. $2x \cos x - x^2 \sin x$

2. $3x^5 \tan x + \frac{1}{2}x^6 \sec^2 x$

3. $-\cot x \, \text{cosec} \, x$

4. $-\text{cosec}^2 x$

5. $\sec x (1 + x \tan x)$

6. $2\cos(4x)$

7. $-6\sin x \cos^5 x$

8. $\sin\left(\frac{x}{2}\right)\cos\left(\frac{x}{2}\right)$

9. $\frac{\sqrt{2}}{6}$

10. $\frac{\pi}{6} - \frac{\sqrt{3}}{8}$

Exercise 2.3 (page 27):

1. e^{7x}

2. $e^{3(2y-x)}$

3. $4x$

4. x^3

5. $2e^{2x}$

6. $x^2 e^{x^3}$

7. $e^x \sin x + e^x \cos x$

8. $\frac{3}{x}$

9. $2xe^{x^2}\ln(4x) + \frac{e^{x^2}}{x}$

10. $12x^2 \ln(x^2) + 8x^2$

Chapter 3

Test Yourself (page 31):

1. $(3, \pi)$

2. $\left(1, \frac{\pi}{4}\right)$

3. $\left(2\sqrt{2}, \frac{3\pi}{4}\right)$

4. $\left(2, \frac{4\pi}{3}\right)$

5. $\theta = \frac{5\pi}{4}$

6. $r^2 - 4r\cos\theta + 3 = 0$

7. $r = 2\sec(\theta - \pi)$

8. $r = a + b\theta$

9. $\left(2\sqrt{3}, 0.955, \frac{7\pi}{4}\right)$

10. $r = 20, t = 50$

Exercise 3.1 (page 36):

1. $(2, 0)$

2. $(4, \frac{3\pi}{2})$

3. $(-2, 0)$

4. $(5\sqrt{2}, 5\sqrt{2})$

5. $(4\sqrt{3}, 4)$

6. $(2\sqrt{2}, \frac{\pi}{4})$

7. $(3, \frac{\pi}{2})$

8. $(2, \frac{5\pi}{6})$

Exercise 3.2 (page 40):

1. $\theta = \frac{\pi}{2}$

2. $\theta = \frac{3\pi}{4}$

3. $\theta = \frac{5\pi}{4}$

4. $r = 4$

5. $r = \frac{1}{\sqrt{2}} \sec(\theta - \frac{3\pi}{4})$

6. $r = 2\sqrt{2} \sec(\theta - \frac{\pi}{4})$

7. $r = 2\cos\theta \quad -\frac{\pi}{2} \le \theta \le \frac{\pi}{2}$

8. $r^2 - 2\sqrt{2}r\cos(\theta - \frac{\pi}{4}) - 47 = 0$

Exercise 3.3 (page 46):

1. $(\sqrt{2}, \frac{\pi}{4}, 1)$

2. $(2\sqrt{2}, \frac{\pi}{4}, \frac{\pi}{2})$

3. $r = 1, \varphi = \frac{\pi}{4}$

4. $r = 4, t = 3$

5. $r = 3\sqrt{2}, \theta = \frac{3\pi}{4}$

6. $r^2 = a\cos(2\theta)$

7. $r^2 = a\sin(2\theta)$

8. $r = a + b\cos\theta$ or $r = a + b\sin\theta$

9. $r = a + b\cos\theta$ or $r = a + b\sin\theta$, and $a = b$

10. An Archimedean spiral

Chapter 4

Test Yourself (page 49):

1. $\mathbb{N}$ is the natural numbers, $\mathbb{Z}$ is the integers, $\mathbb{Q}$ is the rational numbers, $\mathbb{R}$ is the real numbers and $\mathbb{C}$ is the complex numbers.

2. $x = \pm 9i$

3. $3 \pm i$

4. $1 + 5i$

5. $22 - 7i$

6. $\frac{14}{17} - \frac{12}{17}i$

7. $\frac{-7}{10} + \frac{11}{10}i$

8. See Figure C.1 (at end of solutions).

9. $2e^{\frac{\pi}{3}i}$

10. $-8 + 8\sqrt{3}i$

Exercise 4.1 (page 52):

1. α =subtraction

β =division

γ ="Analysis"

$\delta = \sqrt{-1}$

2. -1

3. A complex number is of the form $a + bi$, where a and b are real numbers and $i = \sqrt{-1}$.

4. $\pm 10i$

5. $\pm 8i$

6. $1 \pm i$

7. $-2 \pm 4i$

8. $\frac{1}{4} \pm \frac{1}{4}i$

9. $1 \pm \sqrt{2}i$

10. $\frac{2}{3} \pm \frac{\sqrt{5}}{3}i$

Exercise 4.2 (page 55):

1. $10 + 8i$

2. $7 - i$

3. 0

4. $-3 + 12i$

5. $15 - 9i$

6. $27 + 36i$

7. $23 + 11i$

8. $21 - 15i$

9. $38 + 21i$

10. $76 + 32i$

11. $3 + i$

12. $\frac{5}{2} - \frac{1}{2}i$

13. $\frac{-1}{2} + 2i$

14. $\frac{67}{58} - \frac{37}{58}i$

15. $1 + i$

16. $\frac{3}{2} + \frac{5}{2}i$

Exercise 4.3 (page 59):

1. See Figure C.2 (at end of solutions).

2. $\sqrt{13}$

3. $\sqrt{5}$

4. $\frac{\pi}{2}$

5. $\frac{5\pi}{4}$

6. $2\sqrt{2}e^{\frac{3\pi}{4}i}$

7. $2e^{\frac{5\pi}{6}i}$

8. -324

Chapter 5

Test Yourself (page 61):

1. 10

2. $x = \pm\frac{\sqrt{115}}{12}$

3. $(-7, 19, 9)$

4. $(6\sqrt{17}, 14\sqrt{17}, 2\sqrt{17}, 6\sqrt{17})$

5. $43i + 19j + 15k$

6. 42

7. $(18, 72, 54)$

8. $\theta = \arccos\frac{16}{21}$

9. $(3, -6, 3)$

10. -79

Exercise 5.1 (page 66):

1. 14

2. $\pm\frac{6}{13}$

3. $(1, 14, 5, -5)$

4. $(1, 1, -13, 10, 13)$

5. $(20, 35, 15, 40)$

6. $(-2, 13, 9)$

7. $(-74, 48, 56, -20)$

8. $(63, 58, 23)$

9. $(44, 22, 77)$

10. $(6\sqrt{2}, 27\sqrt{2}, -33\sqrt{2}, 42\sqrt{2})$

Exercise 5.2 (page 69):

1. $2i + 7j + 3k$

2. $u + v = 11i - 11j - 2k$
   $5u = 40i - 10j - 30k$

3. 92

4. 106

5. 87

6. $(158, 79, 316)$

7. $\theta = \arccos\frac{63}{65}$

8. $\theta = \arccos\frac{84}{85}$

Exercise 5.3 (page 71):

1. $(-25, -5, 17)$

2. $(-9, 13, 1)$

3. $(8, 7, -6)$

4. $(16, -2, -26)$

Chapter 6

Test Yourself (page 73):

1. $\begin{pmatrix} 7 & 9 \\ 6 & 0 \end{pmatrix}$

2. $\begin{pmatrix} 26 & -10 \\ 24 & 16 \\ 4 & 17 \end{pmatrix}$

3. $\begin{pmatrix} 2 & 40 \\ -12 & 8 \end{pmatrix}$

4. $\begin{pmatrix} 13 & -6 \\ -1 & 14 \end{pmatrix}$

5. $\begin{pmatrix} 0 & -13 \\ 10 & 22 \\ 32 & 21 \end{pmatrix}$

6. $\begin{pmatrix} 53 & 43 \\ 25 & 6 \\ -3 & 31 \end{pmatrix}$

7. $\begin{pmatrix} -17 & 7 & -14 \\ 28 & 2 & 97 \end{pmatrix}$

8. 11

9. $\begin{pmatrix} 4 & 3 \\ \frac{-5}{2} & -2 \end{pmatrix}$

10. $\begin{pmatrix} -3 & \frac{19}{3} \\ 1 & -2 \end{pmatrix}$

Exercise 6.1 (page 77):

1. $A = (\alpha_{ij})_{3 \times 5}$

2. 3×5

3. $\begin{pmatrix} 7 & 9 \\ 3 & 6 \end{pmatrix}$

4. $\begin{pmatrix} -10 & 7 \\ 3 & 5 \end{pmatrix}$

5. $\begin{pmatrix} 24 & 3 & 12 \\ -6 & 18 & 0 \\ 15 & 3 & -6 \end{pmatrix}$

6. $\begin{pmatrix} -7 & 9 & -8 \\ 10 & 0 & -12 \\ 1 & 15 & 6 \end{pmatrix}$

7. No. B and C are not of the same order, and so addition and subtraction are not possible.

8. $\begin{pmatrix} 1 & 24 \\ 10 & 36 \end{pmatrix}$

9. $\begin{pmatrix} 3 & -3 & 0 \\ -18 & 18 & 51 \end{pmatrix}$

10. $\begin{pmatrix} 6 & 13 & -2 \\ -10 & -13 & -7 \\ 3 & 14 & 15 \end{pmatrix}$

Exercise 6.2 (page 82):

1. $\begin{pmatrix} 2 & 34 \\ 3 & 31 \end{pmatrix}$

2. $\begin{pmatrix} 12 & 11 \\ 20 & 19 \end{pmatrix}$

3. $\begin{pmatrix} 14 & 16 \\ 24 & 39 \\ 14 & 25 \end{pmatrix}$

4. $\begin{pmatrix} -3 & -15 & 8 \\ 16 & 66 & -18 \end{pmatrix}$

5. $\begin{pmatrix} 3 & 4 \\ 2 & 5 \end{pmatrix}$ and $\begin{pmatrix} 3 & 4 \\ 2 & 5 \end{pmatrix}$. They are both the same. It doesn't matter whether we premultiply of postmultiply by the identity matrix, we always get the original matrix back.

6. $\begin{pmatrix} -16 & 6 \\ 0 & 8 \end{pmatrix}$

7. $\begin{pmatrix} 1 & -5 \\ 16 & 43 \end{pmatrix}$

8. $\begin{pmatrix} -6 & -24 \\ 28 & 1 \end{pmatrix}$

9. $\begin{pmatrix} 21 & 4 \\ 5 & 11 \end{pmatrix}$

10. $\begin{pmatrix} 25 & 25 & 25 \\ 0 & 1 & 8 \\ 12 & 13 & 31 \end{pmatrix}$

Exercise 6.3 (page 86):

1. -10

2. -10

3. -9

4. -46

5. A and B are invertible, but C is not.

6. $\begin{pmatrix} \frac{5}{24} & \frac{13}{48} \\ \frac{-1}{8} & \frac{-1}{16} \end{pmatrix}$

7. $\begin{pmatrix} 2 & \frac{7}{3} \\ 1 & 1 \end{pmatrix}$

8. $\begin{pmatrix} \frac{-1}{14} & \frac{-3}{14} \\ \frac{2}{21} & \frac{-1}{21} \end{pmatrix}$

Chapter 7

Test Yourself (page 89):

1. $\begin{pmatrix} -2 & 1 \\ 4 & -2 \end{pmatrix}$

2. $\begin{pmatrix} 0 & -1 \\ 2 & 3 \end{pmatrix}$

3. $x_1 = 3x_0 - 2y_0$, $y_1 = y_0$

4. $\begin{pmatrix} 1 & 0 \\ 0 & -1 \end{pmatrix}$

5. $\begin{pmatrix} \frac{\sqrt{2}}{2} & -\frac{\sqrt{2}}{2} \\ \frac{\sqrt{2}}{2} & \frac{\sqrt{2}}{2} \end{pmatrix}$

6. $\begin{pmatrix} -3 & 0 \\ 0 & -3 \end{pmatrix}$

7. $\lambda = 8$

8. $\lambda = -2$

9. $\lambda = 3, \mathbf{v} = \begin{pmatrix} 1 \\ 1 \end{pmatrix}$ and $\lambda = -2$,

   $\mathbf{v} = \begin{pmatrix} 1 \\ -4 \end{pmatrix}$

10. $\lambda = 1, \mathbf{v} = \begin{pmatrix} 2 \\ 7 \end{pmatrix}$ and $\lambda = -4$,

    $\mathbf{v} = \begin{pmatrix} 1 \\ 1 \end{pmatrix}$

Exercise 7.1 (page 94):

1. The identity map.

2. $\begin{pmatrix} -1 & 0 \\ 0 & 1 \end{pmatrix}$

3. $\begin{pmatrix} 2 & 3 \\ 1 & -4 \end{pmatrix}$

4. $\begin{pmatrix} -3 & 0 \\ 1 & -2 \end{pmatrix}$

5. $x_1 = x_0 + 2y_0$, $y_1 = 5x_0 + 3y_0$

6. $x_1 = 2x_0 - 3y_0$, $y_1 = 6x_0 + 2y_0$

7. $x_1 = 4x_0$, $y_1 = 2x_0 + y_0$

8. $(0,0)$

Exercise 7.2 (page 104):

1. $\begin{pmatrix} \frac{1}{2} & \frac{\sqrt{3}}{2} \\ \frac{\sqrt{3}}{2} & -\frac{1}{2} \end{pmatrix}$

2. $\begin{pmatrix} -\frac{1}{2} & \frac{\sqrt{3}}{2} \\ \frac{\sqrt{3}}{2} & \frac{1}{2} \end{pmatrix}$

3. $\begin{pmatrix} \frac{\sqrt{2}}{2} & -\frac{\sqrt{2}}{2} \\ \frac{\sqrt{2}}{2} & \frac{\sqrt{2}}{2} \end{pmatrix}$

4. $\begin{pmatrix} \frac{1}{2} & -\frac{\sqrt{3}}{2} \\ \frac{\sqrt{3}}{2} & \frac{1}{2} \end{pmatrix}$

5. $\begin{pmatrix} \frac{1}{2} & 0 \\ 0 & \frac{1}{2} \end{pmatrix}$

6. $\begin{pmatrix} 3 & 0 \\ 0 & 6 \end{pmatrix}$

7. $\begin{pmatrix} \frac{3\sqrt{3}}{2} & -\frac{3}{2} \\ \frac{3}{2} & \frac{3\sqrt{3}}{2} \end{pmatrix}$

8. $\begin{pmatrix} 0 & -2 \\ -4 & 0 \end{pmatrix}$

9. $\begin{pmatrix} \frac{\sqrt{3}}{2} & -\frac{1}{2} \\ -\frac{1}{2} & -\frac{\sqrt{3}}{2} \end{pmatrix}$

10. $\begin{pmatrix} 1 & \sqrt{3} \\ \frac{\sqrt{3}}{2} & -\frac{1}{2} \end{pmatrix}$

Exercise 7.3 (page 111):

1. $\lambda = 1$

2. $\lambda = 4$

3. $\lambda = -2$

4. $\lambda = 4$

5. $\lambda = 5, \mathbf{v} = \begin{pmatrix} 3 \\ 1 \end{pmatrix}$ and $\lambda = -2$,

   $\mathbf{v} = \begin{pmatrix} 1 \\ -2 \end{pmatrix}$

6. $\lambda = 5, \mathbf{v} = \begin{pmatrix} -1 \\ 7 \end{pmatrix}$ and $\lambda = -1$,

   $\mathbf{v} = \begin{pmatrix} 1 \\ -1 \end{pmatrix}$

7. $\lambda = 2, \mathbf{v} = \begin{pmatrix} 5 \\ 1 \end{pmatrix}$ and $\lambda = -4$,

   $\mathbf{v} = \begin{pmatrix} 1 \\ -1 \end{pmatrix}$

8. $\lambda = 4, \mathbf{v} = \begin{pmatrix} 1 \\ 3 \end{pmatrix}$ and $\lambda = -6$,

   $\mathbf{v} = \begin{pmatrix} 3 \\ -1 \end{pmatrix}$

9. $\lambda = 3, \mathbf{v} = \begin{pmatrix} 1 \\ 4 \end{pmatrix}$ and $\lambda = 6$,

   $\mathbf{v} = \begin{pmatrix} 1 \\ 1 \end{pmatrix}$

10. $\lambda = -3, \mathbf{v} = \begin{pmatrix} 5 \\ -3 \end{pmatrix}$ and

$\qquad \lambda = -4, \mathbf{v} = \begin{pmatrix} 2 \\ -1 \end{pmatrix}$

Chapter 8

Test Yourself (page 113):

1. $\frac{4x^2+7x+10}{x^2(x+2)}$

2. $\frac{7}{4} \left(\frac{1}{x-2} - \frac{1}{x+2} \right)$

3. $3\ln\left|\frac{x}{x+1}\right| + k$

4. $y = \frac{x}{kx+1}$

5. $y = Ae^{\frac{x^2}{2}}$

6. $y = \sqrt{\frac{1}{2}\ln|xk|}$

7. $y = Ae^{\frac{-1}{9x^3}}$

8. $y = \sqrt{\frac{22}{3}\ln\left|\frac{x}{x+3}\right| + k}$

9. $6\ln\left|\frac{y-1}{y}\right| = x^2 + k$

10. $\frac{1}{2}\ln\left|\frac{y}{y+2}\right| = \ln\left|\frac{x-1}{x+1}\right| + k$

Exercise 8.1 (page 118):

1. $\frac{2x+19}{5x+10}$

2. $\frac{16x+13}{30}$

3. $\frac{2x^3-9}{2x^2-7x+3}$

4. $\frac{-2x}{15}$

5. $\frac{3}{x+2} + \frac{1}{x-4}$

6. $\frac{7}{x-1} + \frac{5}{x+3}$

7. $\frac{6}{x-1} - \frac{5}{x+2}$

8. $\frac{3}{x-2} - \frac{1}{x+2}$

9. $\ln|x| + \ln|x-2| + k$

10. $3\ln|x+1| + 4\ln|x-3| + k$

11. $7\ln|x-4| - \ln|x+1| + k$

12. $3\ln|x-4| + 10\ln|x-5| + k$

Exercise 8.2 (page 123):

1. $y = x^2 + 3x + k$

2. $y = \sqrt{x^2 + k}$

3. $y = kx$

4. $y = Ae^{\frac{x^2}{2}}$

5. $y = \sqrt{2\ln|x| + k}$

6. $y = \sqrt{2x^2 + k}$

Exercise 8.3 (page 126):

1. $x + k = \frac{1}{2}\ln\left|\frac{y-1}{y+1}\right|$

2. $x + k = \ln\left|\frac{y-1}{y}\right|$

3. $y = 2\sqrt{\ln\left|\frac{x-3}{x}\right| + k}$

4. $y = \sqrt{\frac{7}{2}\ln\left|\frac{x-1}{x+1}\right| + k}$

5. $5\ln\left|\frac{y}{y+1}\right| = 2\ln\left|\frac{x-3}{x}\right| + k$

6. $4\ln\left|\frac{y-1}{y+1}\right| = 9\ln\left|\frac{x}{x+1}\right| + k$

Chapter 9

Test Yourself (page 127):

1. $\frac{dy}{dx} + P(x)y = Q(x)$

2. e^{x^4}

3. x^2

4. $e^{x^5}\frac{dy}{dx} + 5x^4 e^{x^5} y = 3e^{x^5} x^2$

5. $\frac{d}{dx}(2xy) = 17x^2$

6. $\frac{d}{dx}(e^{2x^3} y) = 12$

7. $\frac{d}{dx}(e^{x^2} y) = 4x^3 e^{x^2}$

8. $y = 4x + \frac{k}{x}$

9. $y = 5 + ke^{-x^2}$

10. $y = e^x - \frac{e^x}{x} + \frac{k}{x}$

Exercise 9.1 (page 131):

1. $e^{\int P(x)\,dx}$

2. e^{3x}

3. e^{2x^2}

4. $e^{\frac{3x^2}{2}}$

5. $e^{\frac{-1}{2x^2}}$

6. x^2

7. $x\frac{dy}{dx} + y = 6x$

8. $e^{x^3}\frac{dy}{dx} + 3e^{x^3} x^2 y = 4xe^{x^3}$

Exercise 9.2 (page 133):

1. $\frac{d}{dx}(3xy) = 12x$
2. $\frac{d}{dx}(4x^2 y) = e^{x^2}$
3. $\frac{d}{dx}(e^{3x} y) = 16$
4. $\frac{d}{dx}(e^{x^2} y) = 8x$
5. $\frac{d}{dx}(e^{3x^2} y) = 4xe^{3x^2}$
6. $\frac{d}{dx}(x^2 y) = x^4$

Exercise 9.3 (page 135):

1. $y = x + \frac{k}{x}$
2. $y = 1 + \frac{k}{x}$
3. $y = \frac{4}{3} + ke^{-3x}$
4. $y = 3 + ke^{-x^2}$
5. $y = 1 + ke^{-x^4}$
6. $y = 3 - 3e^{-x^3}$

Chapter 10

Test Yourself (page 137):

1. $v = 24, x = 140$
2. $v = \frac{9}{2}, x = 12 + \ln\frac{1}{4}$
3. A body will continue in uniform motion unless acted upon by a force. Force equals rate of change of momentum. For every action, there is an equal and opposite reaction.
4. 15 N
5. 373 N
6. 7 ms^{-1}
7. $\frac{4500}{g}$
8. $T_1 = 2g\sqrt{3}$ and $T_2 = 2g$
9. $\frac{g}{\sqrt{2}}$
10. 30s

Exercise 10.1 (page 144):

1. 28 ms^{-1}
2. 280 m
3. 32 ms^{-1}

4. 44 m
5. $v = 290$ ms$^{-1}, x = 952$ m
6. $v = 8\frac{2}{35}, x = 73\frac{4}{5} + \ln\frac{5}{14}$

Exercise 10.2 (page 148):

1. 30 N
2. 25 ms^{-1}
3. -200 N
4. 9600 N
5. $2\sqrt{2}$ ms^{-1}
6. $20g$
7. $\frac{50}{g}$

Exercise 10.3 (page 154):

1. T
2. $5g\sqrt{2}$
3. $T_1 = \frac{3g\sqrt{3}}{2}, T_2 = \frac{3g}{2}$
4. $T_1 = \frac{8g\sqrt{6}}{\sqrt{3}+1}, T_2 = \frac{16g}{\sqrt{3}+1}$
5. $\sqrt{2}T$
6. $xg\sin y$

Chapter 11

Test Yourself (page 157):

1. "Is not an element of"
2. Yes
3. No
4. 5,15,25.

5.

$\neg$	p
F	T
T	F
1	0

6.

p	$\vee$	q
T	T	T
T	T	F
F	T	T
F	F	F
0	1	0

7. Yes, the statements are logically equivalent.

8. No, the statements are not logically equivalent.

9. None of these.

10. Bijective

Exercise 11.1 (page 161):

1. "Is an element of"

2. $A \setminus B$ is the set of elements in A that are not in B.

3. The set of integer multiples of 4.

4. Yes

5. No

6. 0

7. 5

8. 28

Exercise 11.2 (page 166):

1.

p	$\Leftrightarrow$	q
T	**T**	T
T	**F**	F
F	**F**	T
F	**T**	F
0	1	0

2.

p	$\wedge$	q
T	**T**	T
T	**F**	F
F	**F**	T
F	**F**	F
0	1	0

3. No, the statements are not logically equivalent.

4. Yes, the statements are logically equivalent.

5. No, the statements are not logically equivalent.

6. Yes, the statements are logically equivalent.

Exercise 11.3 (page 169):

1. "For every element in the range,

there is at least one element in the domain that is mapped to it."

2. "There is only one value of x for each value of y" (or, more formally, "No two distinct elements in the domain are mapped to the same element in the range.")

3. No. $a = b \Rightarrow f(a) = f(b)$ is not the correct statement of injectivity (in fact, it is simply a property that every function has).

4. No. The negative part of the range will not be mapped to.

5. Yes

6. No. The negative part of the range will not be mapped to.

7. Whenever an odd number is taken from the domain, there is nowhere for it to be mapped to in the range.

8. Only $\tan x$ is (because $\sin x$ and $\cos x$ don't map outside of the region from -1 to 1).

Chapter 12

Test Yourself (page 171):

1. Every nonempty set of $\mathbb{N}$ has a least element.

2. 1

3. 14

4. 22

5. It works for the first value of n. In our rule, if it's true for n, it's true for $n + 1$.

6. $3 + \cdots + 3n = \frac{3n(n+1)}{2}$

$3 + \cdots + 3n + 3(n + 1)$

$= \frac{3n(n+1)}{2} + 3(n + 1)$

$= \frac{(3n+3)(n+2)}{2}$

$= \frac{(3(n+1))(n+1)+1)}{2}$

7. "Guess" the rule $4n - \frac{n}{2}(n+1)$.

$3 + \cdots + (4-n) = 4n - \frac{n}{2}(n+1)$

$3 + \cdots + (4-n) + (3-n)$

$= 4n - \frac{n}{2}(n+1) + (3-n)$

$= 4n + 4 - \frac{n^2+3n+2}{2}$

$= 4(n+1) - \frac{n+1}{2}(n+2)$

8. I am drinking coffee $\Rightarrow$ I am hot. (*Note:* I am hot $\Rightarrow$ I am drinking coffee is wrong. Nowhere does it say that the only time I get hot is when drinking coffee)

9. The water is not deep here $\Rightarrow$ Diving is not permitted here (Diving is not permitted here $\Rightarrow$ the water is not deep here is wrong)

10. $x > 7 \Rightarrow x \geq 7$. Contrapositive is $x < 7 \Rightarrow x \leq 7$ (You guessed it: any other order is wrong)

Exercise 12.1 (page 173):

1. 1

2. 2

3. 2

4. No. Consider any subset of $\mathbb{N}$ for which we can't find the largest value (like n^2 or n^3) to see that we can't find a greatest element.

5. No. Look at this set: $\{-n|n = 1, 2, 3, \ldots\}$. The numbers in the set are certainly integers, but we will never find the smallest (or "most negative") element.

Exercise 12.2 (page 177):

1. $1 = 1$, so OK.

$\underbrace{1 + 1 + \cdots + 1}_{n \text{ times}} = n$

$\underbrace{1 + 1 + \cdots + 1}_{n \text{ times}} + 1 = n + 1$

So $\underbrace{1 + 1 + \cdots + 1}_{n+1 \text{ times}} = n + 1$

2. $1 + 1 = 2$, so OK.

$2 + \cdots + 2n = n^2 + n$

$2 + \cdots + 2n + 2(n+1)$

$= n^2 + n + 2(n+1)$

$= (n+1)^2 + (n+1)$

3. Recall that RHS $= \left(\frac{n}{2}(n+1)\right)^2$.

$\left(\frac{1}{2}(1+1)\right)^2 = 1$, so OK.

$1^3 + \cdots + n^3 = \left(\frac{n}{2}(n+1)\right)^2$

$1^3 + \cdots + n^3 + (n+1)^3 = \left(\frac{n}{2}(n+1)\right)^2 + (n+1)^3$

$= \frac{(n^2+3n+2)^2}{4}$

$= \left(\frac{n+1}{2}(n+2)\right)^2$

4. "Guess" the rule n^2.

$1 = 1$, so OK.

$1 + \cdots + (2n-1) = n^2$

$1 + \cdots + (2n-1) + (2(n+1)-1)$

$= n^2 + (2(n+1)-1)$

$= (n+1)^2$

5. "Guess" the rule $2n(n+1)$.

$2(1+1) = 4$, so OK.

$4 + \cdots + 4n = 2n(n+1)$

$4 + \cdots + 4n + 4(n+1)$

$= 2n(n+1) + 4(n+1)$

$= (2n+2)(n+2)$

$= (2(n+1))(n+2)$

Exercise 12.3 (page 180):

1. • I have been up the Eiffel Tower at least once $\Rightarrow$ I don't have a phobia of heights.

• x is not odd $\Rightarrow$ either x is 2 or x is not prime.

• The positive number is not divisible by 3 $\Rightarrow$ The sum of the digits is not divisible by 3.

• $x \leq 0 \Rightarrow x$ is not positive.

2. If the number is a positive whole number, we see from *iv* that $x > 0$. It is therefore either odd or even. Now, the contrapositive of *ii* is helpful: if the number we choose

is not odd, and is not 2, then we see for sure that the number is not prime.

Chapter 13

Test Yourself (page 183):

1. 362880
2. 1
3. $\frac{10!}{4!6!}$
4. $\frac{n!}{r!(n-r)!}$
5. $25\left(\frac{24}{25}\right)^{20}$
6. $\dfrac{\binom{13}{1}\binom{13}{2}\binom{13}{3}\binom{13}{4}}{\binom{52}{10}}$
7. 0.94
8. $\frac{1}{3}$
9. $\frac{19}{99}$
10. $\frac{4}{11}$

Exercise 13.1 (page 187):

1. 120
2. 40320
3. 10
4. 15
5. 8
6. $\frac{n!}{r!(n-r)!}$
7. $\frac{1}{3}$
8. $\frac{1}{2}$
9. Yes
10. No

Exercise 13.2 (page 190):

1. $20\left(\frac{19}{20}\right)^{20}$
2. $10\left(\frac{9}{10}\right)^{20}$
3. $\left(\frac{1}{6}\right)^{10}$
4. $\binom{10}{8}\left(\frac{1}{6}\right)^{10}$

5. 0.72
6. £8.70

Exercise 13.3 (page 193):

1. $\frac{1}{4}$
2. 0
3. $\frac{1}{3}$
4. $\frac{1}{8}$
5. $\frac{1}{4}$

Chapter 14

Test Yourself (page 195):

1. Poisson
2. Binomial
3. $\binom{10}{4}\left(\frac{1}{13}\right)^4\left(\frac{12}{13}\right)^6$
4. $\binom{5}{2}\left(\frac{2}{9}\right)^2\left(\frac{7}{9}\right)^3$
5. $\frac{e^{-250}250^{250}}{250!}$
6. $\frac{e^{-10}10^5}{5!}$
7. $\mu = 2$, Var= 1.6
8. $\mu = 5$, Var= 5
9. Use Poisson approximation. $\frac{e^{-5}5^6}{6!}$
10. Use Poisson approximation. $1 - \left(4e^{-3} + \frac{9e^{-3}}{2}\right)$

Exercise 14.1 (page 198):

1. Bernoulli
2. Binomial
3. $\left(\frac{1}{2}\right)^{10}$
4. $\binom{20}{10}\left(\frac{1}{2}\right)^{20}$
5. $\left(\frac{3}{4}\right)^{25} + \frac{25}{4}\left(\frac{3}{4}\right)^{24}$
6. The probability of choosing a particular colour in a specific trial changes because it is dependent on the outcomes of the previous trials.

Exercise 14.2 (page 201):

1. Every event is independent. The average rate does not change over the interval we examine. The number of events that occur in a given time period depends only on the length of the time period and the average rate of occurrances. For a tiny time interval, either 0 or 1 events will occur (i.e., two or more events cannot occur at exactly the same instant).

2. $X \sim Po(7)$

3. $\frac{25}{2}e^{-5}$

4. $\frac{e^{-45}45^{60}}{60!}$

5. $\frac{e^{-36}36^{40}}{40!}$

6. $e^{-0.2}$

7. $3e^{-3}$

8. $\frac{e^{-10}10^{15}}{15!}$

Exercise 14.3 (page 204):

1. 0.6

2. 1.2

3. 5

4. 12

5. "n is large and p is small"

6. It is difficult to calculate the value of the "choose" term with very large numbers, and this isn't necessary when working with Poisson.

7. $\frac{e^{-4}4^6}{6!}$

8. $1 - 3e^{-2}$

Chapter 15

Test Yourself (page 207):

1. The behavioural approach to probability asks an individual which of two bets they prefer, then changes one of the bets to make it more or less favourable. This process is repeated until the individual being asked is indifferent between the two bets, at which point the individual's personal probability elicitation can by found by finding the probability of success of the bet that was being modified. This is a better approach than the frequency approach because it does not require that we repeat an experiment "many times," as we do not know the answer to the question, "How many is many?"

2. It would have $\frac{1}{4}$ (i.e., 90 degrees) coloured the "success" colour, and $\frac{3}{4}$ (i.e., 270 degrees) coloured the "failure" colour.

3. 70%

4. 60 degrees

5. They value reward at its exact monetary value (so they are adopting a "1 for 1" utility function).

6. £2.00

7. £1.50

8. $\frac{5}{42}$

9. Buy 500 plants

10. If the probability of business being good is greater than $\frac{3}{7}$, then go to recruitment agency A, otherwise go to recruitment agency B.

Exercise 15.1 (page 211):

1. The probability of an event happening is equal to the number of times that it happened, divided by the number of trials.

2. The frequency approach is only valid if we repeat the trial "many times" – but how many is many?

3. When the number of trials tends to infinity, the number of successes divided by the total number of trials will tend towards the probability of a single success, with probability 1.

4. $\frac{1}{2}$

5. 108 degrees

6. $\frac{3}{20}$

Exercise 15.2 (page 214):

1. A utility function is a function that assigns a value to a reward.

2. £2.50

3. $x > 20p$

4. 24p

5. $\frac{1}{50}$

6. $\frac{1}{40}$

7. $x > \frac{25}{9}$

Exercise 15.3 (page 219):

1. Buy deluxe regardless of the probability of it being a good year.

2. Buy 1000.

3. If they believe that the probability of the stationery shop being popular is greater than $\frac{2}{15}$, they should go into the large unit, otherwise they should go into the small unit.

Chapter 16

Test Yourself (page 221):

1. Yes

2. Yes

3. Yes

4. Maybe: we cannot deduce for certain either "yes" or "no."

5. $x = \frac{3\pi}{8}$

6. There are a number of different ways to prove this, so as long as your proof is rigorous and unambiguous, that's fine. For the proof used in the chapter, see the part of section 16.2 entitled "An Old Friend" (on page 227).

7. $x = 2\sqrt{2}$

8. Again, there are a number of different ways to prove this, so as long as your proof is rigorous and unambiguous, that's fine. For the proof used in the chapter, see the part of section 16.2 entitled "The Sine Rule" (on page 226).

9. It must be $\frac{3\pi}{2}$. By restricting the length of the lines to the distance between the equator and the North Pole, our lines are each $\frac{1}{4}$ of the length of an equator. This yields the angle sum of $\frac{3\pi}{2}$.

10. If we call the sum of the angles x, then $\pi < x < 3\pi$.

Exercise 16.1 (page 225):

1. Yes

2. No

3. No

4. 1 (the centre of enlargement is unimportant when the scale factor is 1).

5. a) Congruent
    b) Not congruent
    c) Can't say
    d) Congruent
    e) Congruent
    f) Can't say

Exercise 16.2 (page 233):

1. $\frac{2\pi}{7}$

2. a and d

3. d

4. $\frac{b}{2}$

5. From iv, $OC \sin C\hat{O}D = \frac{b}{2}$.
   $OC = R$, $C\hat{O}D = C\hat{B}A$ (because the angle at the centre is twice the angle at the edge).
   So $R \sin B = \frac{b}{2}$, and hence $\frac{b}{\sin B} = 2R$.

6. Refer to figure C.3 (at end of solutions).
   $OA \sin A\hat{O}D = \frac{c}{2}$
   $OA = R$
   $A\hat{O}D = C$
   $R \sin C = \frac{c}{2}$
   $\frac{c}{\sin C} = 2R$.

Exercise 16.3 (page 236):

1. We can't say. Depending on the lengths of the sides, it varies.

2. π. Two lines from the equator to the north pole that are coincident have $\frac{\pi}{2}$ at each side of the equator, and 0 at the north pole, summing in total to π.

3. 3π. Two lines from the equator to the north pole that are coincident have $\frac{\pi}{2}$ at each side of the equator, and *when traversing the reflex angle*, 2π at the north pole, summing in total to 3π.

4. Simply divide any polygon up so that it is made of lots of triangles! This is possible for every polygon.

5. It must be a straight line through the centre of the sphere.

6. It must be a plane (flat surface) that passes through the centre of the sphere.

7. 6000 km

8. π m

Chapter 17

Test Yourself (page 241):

1. $\frac{e^6+1}{e^6-1}$

2. $\ln(2 \pm \sqrt{3})$

3. $\ln(\sqrt{2})$

4. $8\cosh(4x)$

5. $3\coth(3x)$

6. $2e^{2x}\sinh x + e^{2x}\cosh x$

7. $e^{\tanh x}\mathrm{sech}^2 x$

8. $\frac{1}{3}\tanh^3 x + c$

9. $\frac{2}{3}\sinh(3x) + c$

10. $\frac{1}{12}\sinh(6x) - \frac{x}{2} + c$

Exercise 17.1 (page 246):

1. $\frac{e^5-e^{-5}}{2}$

2. $\frac{e^{\frac{3}{2}}+e^{\frac{-3}{2}}}{2}$

3. $\frac{e^2-e^{-2}}{e^2+e^{-2}}$

4. $\frac{e^{\sqrt{2}}+e^{-\sqrt{2}}}{e^{\sqrt{2}}-e^{-\sqrt{2}}}$

5. $x = \ln(3 + \sqrt{10})$

6. $x = \ln(4 \pm \sqrt{15})$

7. $x = \ln(\sqrt{3})$

8. $x = \ln\sqrt{\frac{7}{5}}$

Exercise 17.2 (page 250):

1. RHS$= \frac{1}{4}e^{x+y} + \frac{1}{4}e^{x-y} + \frac{1}{4}e^{y-x}$
   $+ \frac{1}{4}e^{-(x+y)} + \frac{1}{4}e^{x+y} - \frac{1}{4}e^{x-y}$
   $- \frac{1}{4}e^{y-x} + \frac{1}{4}e^{-(x+y)}$
   $= \frac{1}{2}\left(e^{x+y} + e^{-(x+y)}\right)$
   $= \cosh(x+y)$.

2. $\frac{2\sinh x\cosh x}{\cosh^2 x+\sinh^2 x}$
   Divide top and bottom by $\cosh^2 x$:
   $\frac{2\tanh x}{1+\tanh^2 x}$

3. a) $3\cosh(3x)$
   b) $-2\,\mathrm{sech}\,x\tanh x$
   c) $e^x\sinh x + e^x\cosh x$

d) $2x \operatorname{sech}^2 x^2$

e) $\operatorname{cosech}(\ln x)$
   $-\operatorname{cosech}(\ln x) \coth(\ln x)$

f) $\tanh x$

g) $e^{\sinh x} \cosh x$

h) $\coth x \cos x - \operatorname{cosech}^2 x \sin x$

i) $e^{\cosh x} - \operatorname{cosech} x \coth x e^{\cosh x}$

Exercise 17.3 (page 251):

1. $\frac{1}{2}\sinh(2x) + c$

2. $\frac{1}{2}\cosh(4x) + c$

3. $\frac{1}{2}e^{2x} + x + c$

4. $e^x + e^{-x} + c$

5. $3\tanh(3x) + c$

6. $\frac{1}{2}\sinh(2x) + x + c$

7. $\frac{1}{3}\sinh^3 x + c$

8. $\frac{-1}{3}\operatorname{sech}(3x) + c$

9. $\tanh^4 x + c$

Chapter 18

Test Yourself (page 253):

1. $k = mg\tan\theta$

2. $\left(25\sqrt{3} + 5 - \frac{3\sqrt{2}g}{5}\right)\mathrm{ms}^{-2}$

3. $300\mathrm{ms}^{-2}$ towards the centre of the circle.

4. $32\ \mathrm{ms}^{-2}$

5. $8\ \mathrm{N}$

6. $45\ \mathrm{N}$

7. $C = \{(t, 3t + 2)|t \in \mathbb{R}\}$

8. $C = \{(t, \sin t)|t \in \mathbb{R}\}$

9. $15\sqrt{2}$

10. $\frac{99}{20}$

Exercise 18.1 (page 260):

1. $\frac{g\sqrt{3}}{2} - \frac{g}{5}$

2. $\frac{g}{2\sqrt{2}}$

3. Down

4. Up

5. $\left(\frac{\sqrt{2}g}{4} + \frac{1}{2} - \sqrt{3}\right)\mathrm{ms}^{-2}$
   down the slope.

6. $\frac{10g}{\sqrt{3}}$

Exercise 18.2 (page 267):

1. Towards the centre of the circle.

2. $5\pi\ \mathrm{ms}^{-1}$

3. $2\ \mathrm{ms}^{-2}$

4. $45\ \mathrm{ms}^{-2}$

5. $36\ \mathrm{N}$

6. $20\ \mathrm{N}$

Exercise 18.3 (page 271):

1. $\{(t, t)|t \in \mathbb{R}\}$

2. $\{(t, t^2)|t \in \mathbb{R}\}$

3. $\{(3 + 2\cos\theta, 3 + 2\sin\theta)|\theta \in [0, 2\pi)\}$

4. $\{(2 + 2\cos\theta, 2\sin\theta)|\theta \in [0, \pi]\}$

5. 52

6. 80

7. $\int_0^5 \sqrt{1 + 9t^4}\,dt$

8. $\int_0^3 \sqrt{1 + 16t^2}\,dt$

Chapter 19

Test Yourself (page 273):

1. See Figure C.4 (at end of solutions). Joining the dots is *wrong*.

2. See Figure C.5 (at end of solutions). Joining the dots is *wrong*.

3. Yes it is (it is a strictly increasing, and therefore increasing, sequence).

4. Yes

5. No

6. No

7. Yes

8. 0

9. Neither

10. ∞

Exercise 19.1 (page 276):

1. See Figure C.6 (at end of solutions). Joining the dots is *wrong*.
2. See Figure C.7 (at end of solutions). Joining the dots is *wrong*.
3. See Figure C.8 (at end of solutions). Joining the dots is *wrong*.
4. See Figure C.9 (at end of solutions). Joining the dots is *wrong*.
5. a) Increasing, strictly increasing, monotonic
    b) Decreasing, monotonic
    c) Non-monotonic
    d) Non-monotonic

Exercise 19.2 (page 280):

1. Yes. We cannot conclude anything from this fact alone. We need to test whether there's a point after which the sequence is *always* greater than this value, and then whether this still holds if we choose *any* number, not just 12.
2. No. We can conclude for certain that $a_n = -n$ does not tend to ∞.
3. Yes. We cannot conclude anything from this fact alone. We need to test whether this still holds if we choose *any* number, not just 42.
4. Although some individual terms are greater than 20, there is no point after which *every* term is. We can conclude for certain that $a_n = (-2)^n$ *does not* tend to ∞.
5. Yes
6. Yes
7. No
8. Yes

Exercise 19.3 (page 282):

1. Yes (at the beginning of the sequence). We cannot conclude anything from this.
2. No. We can conclude for certain that $a_n = 3n$ does not tend to 0.
3. Yes. We cannot conclude anything from this.
4. No. We can conclude for certain that $a_n = (-3)^n$ does not tend to 0.
5. Yes
6. No
7. No
8. Yes

Chapter 20

Test Yourself (page 285):

1. $n^2 + n$
2. No
3. 0
4. The sequence tends to 0.
5. We can always find a finite number of terms that sum to at least 1, then of the remaining terms we can always find a finite number of terms that sum to at least 1, then of the remaining terms we can always find a finite number of terms that sum to at least 1, ...
6. The first pie has $\frac{1}{3}$ filled in, the second has $\frac{1}{3} + \frac{1}{9} = \frac{4}{9}$ filled in, the third has $\frac{1}{3} + \frac{1}{9} + \frac{1}{27} = \frac{13}{27}$ filled in, and so on. Visually, it is clear that there will never be more than $\frac{1}{2}$ a pie, and so a finite value exists for the summation of infinitely many terms.
7. No
8. Yes

9. $\frac{1}{2^n}$ does, and every term in this sequence is simply 0.2 times the corresponding term in $a_n = \frac{1}{2^n}$.

10. $\frac{1}{n}$ does not, and every term in this sequence is simply 0.3 times the corresponding term in $a_n = \frac{1}{n}$.

Exercise 20.1 (page 288):

1. $\frac{n}{6}(n+1)(2n+1)$
2. n^2
3. No
4. No
5. 0
6. 1
7. No
8. Yes

Exercise 20.2 (page 291):

1. The sequence $a_n = \frac{1}{n}$.

2. No

3. No

4. Yes

5. $\frac{2}{3}$

6. Yes

Exercise 20.3 (page 293):

1. $a_n \le K \cdot b_n$, for all n
   $\sum_{n=1}^{\infty} b_n$ is a finite number.

2. K is a positive, real number.

3. No (we can only use it on non-negative sequences).

4. Yes

5. Compare to $b_n = \frac{1}{2^n}$ to see that $a_n = \frac{3}{2^n}$ does have a finite value for its infinite sum.

6. Compare to $b_n = 1$ to see that $a_n = 1 + \frac{3}{2^n}$ does not have a finite value for its infinite sum.

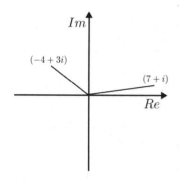

Figure C.1

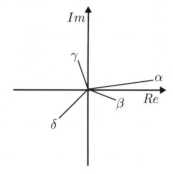

Figure C.2

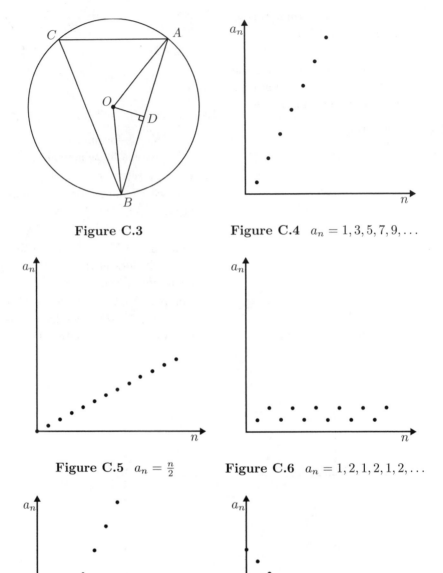

Figure C.3

Figure C.4 $a_n = 1, 3, 5, 7, 9, \ldots$

Figure C.5 $a_n = \frac{n}{2}$

Figure C.6 $a_n = 1, 2, 1, 2, 1, 2, \ldots$

Figure C.7 $a_n = 2, 4, 6, 8, \ldots$

Figure C.8 $a_n = 4 - n$

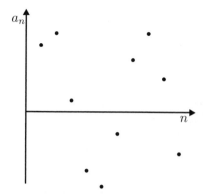

Figure C.9 $a_n = \sin n$

Index

Printed in the United States